George Bruce Halsted

Metrical geometry

An elementary treatise on mensuration

George Bruce Halsted

Metrical geometry
An elementary treatise on mensuration

ISBN/EAN: 9783742892317

Manufactured in Europe, USA, Canada, Australia, Japa

Cover: Foto ©berggeist007 / pixelio.de

Manufactured and distributed by brebook publishing software (www.brebook.com)

George Bruce Halsted

Metrical geometry

AN ELEMENTARY TREATISE

ON

MENSURATION

BY

GEORGE BRUCE HALSTED

A.B., A.M., AND EX-FELLOW OF PRINCETON COLLEGE; PH.D. AND
EX-FELLOW OF JOHNS HOPKINS UNIVERSITY; INSTRUCTOR IN
POST-GRADUATE MATHEMATICS, PRINCETON COLLEGE.

SECOND EDITION.

BOSTON:
PUBLISHED BY GINN, HEATH, & CO.
1881.

INSCRIBED TO

J. J. SYLVESTER,

A.M., Cum.; F.R.S., L. and E.; Cor. Mem. Institute of France; Mcm. Acad. of Sciences in Berlin, Göttingen, Naples, Milan, St. Petersburg, etc.; LL.D., Univ. of Dublin, U. of E.; D.C.L., Oxford; Hon. Fellow of St. John's Col., Cam.,

IN TOKEN OF THE INESTIMABLE BENEFITS DERIVED FROM TWO YEARS' WORK WITH HIM,

BY THE AUTHOR.

PREFACE.

THIS book is primarily the outcome of work on the subject while teaching it to large classes.

The competent critic may recognize signs of a Berlin residence; but a considerable part, it is believed, is entirely new.

Special mention must be made of the book's indebtedness to Dr. J. W. DAVIS, a classmate with me at Columbia School of Mines; also to Prof. G. A. WENTWORTH, who has kindly looked over the proofs. But if the book be found especially accurate, this is due to the painstaking care of my friend H. B. FINE, Fellow of Princeton.

Any corrections or suggestions relating to the work will be thankfully received.

GEORGE BRUCE HALSTED.

PRINCETON, NEW JERSEY,
May 12, 1881.

CONTENTS.

MENSURATION.

	Page.
INTRODUCTION	1
THE METRIC SYSTEM	2
NOTATION AND ABBREVIATIONS	3

CHAPTER I.

THE MEASUREMENT OF LINES.

§ (A).—STRAIGHT LINES.

ILLUSTRATIVE PROBLEMS	5
(α) To measure a line the ends of which only are accessible	5
(β) To find the distance between two objects, one of which is inaccessible	5
(γ) To measure a line when both ends of it are inaccessible	6
(δ) To measure a line wholly inaccessible	6

§ (B).—STRAIGHT LINES IN TRIANGLES.

Article.
I. RIGHT-ANGLED TRIANGLES	6
1. To find hypothenuse	6
2. To find side	7
II. OBLIQUE TRIANGLES	7
3. Obtuse	8
4. Acute	8
5. Given perpendicular	9
6. To find medials	9
III. STRAIGHT LINES IN SIMILAR FIGURES	10
7. To find corresponding line	10
IV. CHORDS OF A CIRCLE	10
8. To find diameter	10

Article.	Page.
9. To find height of arc	11
10. Given chord and height	12
11. Given chord and radius, to find chord of half the arc	12
12. To find circumscribed polygon	13

§ (C). — METHOD OF LIMITS.

V. Definition of a Limit	14
13. Principle of Limits	15
14. The length of the curve	16

§ (D). — THE RECTIFICATION OF THE CIRCLE.

15. Length of semicircumference	19
16. Circumferences are as their radii	20
VI. Lines in any Circle	21
17. Value of π	21
18. To find circumference	21
19. Value of diameter	21

CHAPTER II.

The Measurement of Angles.

§ (E). — THE NATURAL UNIT OF ANGLE.

VII. Angles are as Arcs	22
20. Numerical measure of angle	22
VIII. Angle Measured by Arc	23

§ (F). CIRCULAR MEASURE OF AN ANGLE.

21. Angle in radians	24
22. To find length of arc	24
23. To find number of degrees in arc	25
IX. Arcs which Correspond to Angles	25
24. Angle at center	25
25. Inscribed angle	26
26. Angle formed by tangent and chord	26
27. Angle formed by two chords	26
28. Secants and tangents	26

Article.		Page.
29.	Given degrees to find circular measure	26
30.	Given circular measure to find degrees	27
31.	Given angle and arc to find radius	27
	X. ABBREVIATIONS FOR AREAS	28

CHAPTER III.

THE MEASUREMENT OF PLANE AREAS.

§ (G).—PLANE RECTILINEAR FIGURES.

	XI. MEASURING AREA OF SURFACE	29
32.	Area of rectangle	29
33.	Area of square	32
34.	Area of parallelogram	33
35.	Area of triangle (given altitude)	33
36.	Area of triangle (given sides)	34
37.	Radius of inscribed circle	35
38.	Radius of circumscribed circle	36
39.	Radius of escribed circle	37
	XII. TRAPEZOID AND TRIANGLE AS TRAPEZOID	38
40.	Area of trapezoid	38
	XIII. COÖRDINATES OF A POINT	39
	XIV. POLYGON AS SUM OF TRAPEZOIDS	40
41.	To find sum of trapezoids	40
42.	Area of any polygon	43
43.	Area of quadrilateral	45
44.	Area of a similar figure	48
	XV. CONGRUENT AND EQUIVALENT	48
	XVI. PROPERTIES OF REGULAR POLYGON	49
45.	Area of regular polygon	49
46.	Table of regular polygons	50

§ (H).—AREAS OF PLANE CURVILINEAR FIGURES.

47.	Area of circle	51
48.	Area of sector	51
49.	Area of segment	52
50.	Circular zone	54
51.	Crescent	54

Article.	Page.
52. Area of annulus	54
53. Area of sector of annulus	55
XVII. CONICS	56
54. Area of parabola	57
55. Area of ellipse	59

CHAPTER IV.

MEASUREMENT OF SURFACES.

XVIII. DEFINITIONS RELATING TO POLYHEDRONS	61
56. Faces plus summits exceed edges by two	61

§ (I).—PRISM AND CYLINDER.

XIX. PARALLELEPIPED AND NORMAL	62
57. Mantel of prism	63
XX. CYLINDRIC AND TRUNCATED	64
58. Mantel of Cylinder	64

§ (J).—PYRAMID AND CONE.

XXI. CONICAL AND FRUSTUM	66
59. Mantel of pyramid	66
60. Mantel of cone	67
61. Mantel of frustum of pyramid	68
62. Mantel of frustum of cone	69
63. Frustum of cone of revolution	70

§ (K).—THE SPHERE.

XXII. SPHERE A SURFACE, GLOBE A SOLID	71
64. Area of a sphere	71
XXIII. SPHERICAL SEGMENT AND ZONE	73
65. Area of a zone	73
66. Theorem of Pappus	74
67. Surface of solid ring	74

§ (L).—SPHERICS AND SOLID ANGLES.

Article.	Page.
XXIV. STEREGON AND STERADIAN	77
68. Area of a lune	78
XXV. SOLID ANGLE MADE BY TWO, THREE, OR MORE PLANES	79
XXVI. SPHERICAL PYRAMID	80
69. Solid angles are as spherical polygons	80
XXVII. SPHERICAL EXCESS	80
70. Area of spherical triangle	80
71. Area of spherical polygon	·81
TABLE OF ABBREVIATIONS	83

CHAPTER V.

THE MEASUREMENT OF VOLUMES.

§ (M).—PRISM AND CYLINDER.

XXVIII. SYMMETRICAL AND QUADER	84
XXIX. VOLUME AND UNIT OF VOLUME	84
XXX. LENGTHS, AREAS, AND VOLUMES ARE RATIOS	84
72. Volume of quader	85
XXXI. MASS, DENSITY, WEIGHT	86
73. To find density	87
74. Volume of parallelepiped	88
75. Volume of prism	90
76. Volume of cylinder	91
77. Volume of cylindric shell	92

§ (N).—PYRAMID AND CONE.

XXXII. ALTITUDE OF PYRAMID	93
78. Parallel sections of pyramid	93
79. Equivalent tetrahedra	94
80. Volume of pyramid	95
81. Volume of cone	97

§ (O).—PRISMATOID.

Article.	Page.
XXXIII.–XLI. Definitions Respecting Prismatoid,	98–101
82. Volume of prismatoid	101
83. Volume of frustum of pyramid	104
84. Volume of frustum of cone	106
85. Volume of ruled surface	107
86. Volume of wedge	108
87. Volume of tetrahedron	109

§ (P). SPHERE.

88. Volume of sphere	110
89. Volume of spherical segment	112
XLII. Generation of Spherical Sector	113
90. Volume of spherical sector	113
XLIII. Definition of Spherical Ungula	114
91. Volume of spherical ungula	114
92. Volume of spherical pyramid	115

§ (Q).—THEOREM OF PAPPUS.

93. Theorem of Pappus	115
94. Volume of ring	116

§ (R).—SIMILAR SOLIDS.

XLIV. Similar Polyhedrons	118
95. Volume of similar solids	118

§ (S).—IRREGULAR SOLIDS.

96. By covering with liquid	119
97. Volume by weighing	120
98. Volume of irregular polyhedron	120

CHAPTER VI.

The Applicability of the Prismoidal Formula.

99. Test for applicability	121

§ (T). — PRISMOIDAL SOLIDS OF REVOLUTION.

Article   Page
XLV. Examination of the Different Cases . . 124

§ (U). — PRISMOIDAL SOLIDS NOT OF REVOLUTION.

XLVI. Discussion of Cases 127

§ (V). — ELIMINATION OF ONE BASE.

100. Prismatoid determined from one base . . 130

CHAPTER VII.

Approximation to all Surfaces and Solids.

§ (W). — WEDDLE'S METHOD.

101. Seven equidistant sections 131

CHAPTER VIII.

Mass-Center.

§ (X). — FOR HOMOGENEOUS BODIES.

102–104. Introductory 136
    XLVII. By Symmetry 136
105. Definition of symmetric points 136
106–117. Direct μC deductions 137
118. Triangle 138
119. Perimeter of triangle 138
120–122. Tetrahedron 138
123. Pyramid and cone 138
    XLVIII. The Mass-Center of a Quadrilateral . 139
124. Definition of a sect 139
125. Of an opposite on a sect 139
126. μC by opposites 139
127. Geometric method of centering quadrilateral . . . 139
128. The mass-center of an octahedron 140

Article.	Page.
XLIX. GENERAL MASS-CENTER FORMULA . . . 141	
129. Mass-center of any prismatoid 142	
130. Test for applicability 142	
131. Average haul 144	
132. The $^\mu C$ of a consecutive series 144	

EXERCISES AND PROBLEMS IN MENSURATION.

[These are arranged and classed in accordance with the above 132 Articles] 145

LOGARITHMS 225

A TREATISE ON MENSURATION.

AN ELEMENTARY

TREATISE ON MENSURATION.

INTRODUCTION.

MENSURATION is that branch of mathematics which has for its object the measurement of geometrical magnitudes. It has been called, that branch of applied geometry which gives rules for finding the length of lines, the area of surfaces, and the volume of solids, from certain data of lines and angles.

A **Magnitude** is anything which can be conceived of as added to itself, or of which we can form multiples.

The **measurement** of a magnitude consists in finding how many times it contains another magnitude of the same kind, taken as a unit of measure. Measurement, then, is the process of ascertaining the ratio which one magnitude bears to some other chosen as the standard; and *the measure* of a magnitude is this ratio expressed in numbers. Hence, we must refer to some concrete standard, some actual object, to give our measures their absolute meaning.

The concrete standard is arbitrary in point of theory, and its selection a question of practical convenience.

A discrete aggregate, such as a pile of cannon-balls, or a number, has a natural unit, — "one of them."

But in the continuous quantity, space, with which we chiefly have to deal, the fundamental unit, a length, is defined by fixing upon a physical object, such as a bar of platinum, and agreeing to refer to its length as our standard. That is, we assume some arbitrary length in terms of which all space measurements are to be expressed.

The one actually adopted is the **Meter**, which is the length of a special bar deposited in the French archives. This we choose because of the advantages of the metric system, which applies only a decimal arithmetic, and has a uniform and significant terminology to indicate the multiples and submultiples of a unit.

THE METRIC SYSTEM

convenes to designate multiples by the Greek numerals, and submultiples by the corresponding Latin words; as follows:

Multiple.	Name.	Derivation.	Meaning.
		Greek.	
10,000	myria-	μυριάς.	ten thousand.
1,000	kilo-	χίλιοι.	a thousand.
100	hecto-	ἑκατόν.	a hundred.
10	deka-	δέκα.	ten.
Divisions.		*Latin.*	
$\frac{1}{10}$	deci-	decem.	ten.
$\frac{1}{100}$	centi-	centum.	a hundred.
$\frac{1}{1000}$	milli-	mille.	a thousand.

So a millimeter (^{mm}) is one-thousandth of a meter.

Thus we are given a number of subsidiary units. For any particular class of measurements, the most convenient of these may be chosen.

The kilometer (^{km}) is used as the unit of distance; and along roads and railways are placed kilometric poles or stones.

The centimeter (^{cm}) is the most common submultiple of the meter.

A chief advantage of this decimal system of measures is, that in it reduction involves merely a shifting of the decimal point.

NOTATION AND ABBREVIATIONS
TO BE USED IN THIS BOOK.

Large letters indicate points; thus, A, B, and C denote the three angular points of a triangle, or C may denote the center of a circle, while A and B are on the circumference; then AB will denote the chord joining A to B; and, generally, AD means a straight line terminated at A and D.

In the formulæ, small letters are used to denote the numerical measures of lines; so that ab, as in common algebra, denotes the product of two numbers.

The following choice of letters is made for writing a formula:

[TABLE FOR REFERENCE.]

a, b, and c are the sides of any triangle, respectively, opposite the angular points A, B, C.

If the triangle is right-angled, $a =$ altitude, $b =$ base, $c =$ hypothenuse.

In regard to a circle, $c =$ circumference, $d =$ diameter.

$e =$ spherical excess.
$f =$ flat angle.
$g =$ number of degrees in an angle or arc.
$g°$ means expressed in degrees only, g' minutes, g'' seconds.
$h =$ height.
$i =$ medial line.

$j =$ projection.
$k =$ chord.
$l =$ length.
$m =$ meter.
$n =$ any number.
$o =$ perigon.
$p =$ perimeter.
$q =$ square.
$r =$ radius.
$s = \frac{1}{2}(a + b + c)$.
$t =$ tangent.
$u =$ circular measure.
$v =$ a discrete variable.
$w =$ width.
$\left.\begin{array}{l} x \\ y \\ z \end{array}\right\} =$ coördinates of a point.

CHAPTER I.

THE MEASUREMENT OF LINES.

§ (A).—STRAIGHT LINES.

An accessible straight line is practically measured by the direct application of a standard length suitably divided.

If the straight line contain the standard unit n times, then n is its numerical measure.

But, properly speaking, any description of a length by counting of standard lengths is imperfect and merely approximate. Few physical measurements of any kind are exact to more than six figures, and that degree of accuracy is very seldom obtainable, even by the most delicate instruments. Thus, in comparing a particular meter with the standard bar, a difference of a thousandth of a millimeter can be detected.

In four measurements of a base line at Cape Comorin, it is said the greatest error was 0.077 inch in 1.68 mile, or one part in 1,382,400; and this is called an almost incredible degree of accuracy.

When we only desire rough results, we may readily shift the place of the line to be measured, so as to avoid natural obstacles. Still, under the most favorable circumstances, all actual measurements of continuous quantity are only approximately true. But such imperfections, with the devised methods of correction, have reference to the physical measurement of things; to the data, then, which in book-questions we suppose accurately given.

ILLUSTRATIVE PROBLEMS.

(α) *To measure a line the ends of which only are accessible.*

Suppose AB the line. Choose a point C from which A and B are both visible. Measure AC, and prolong it until $CD = AC$. Measure BC, and prolong it until $CE = BC$. Then $ED = AB$.

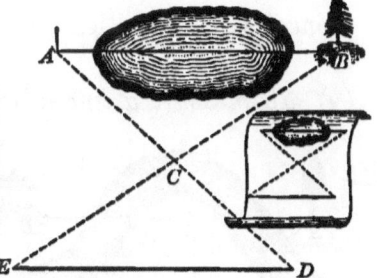

Ww. 49 & 106; (Eu. I. 15 & 4; Cv. I. 23 & 76).

NOTE. *Ww.* refers to Wentworth's Geometry, third edition, 1879. *Eu.* refers to Todhunter's Euclid, new edition, 1879. *Cv.* refers to Chauvenet's Geometry. These parallel references are inserted in the text for the convenience of students having either one of these geometries at hand. References to preceding parts of this Mensuration will give simply the number of the article.

(β) *To find the distance between two objects, one of which is inaccessible.*

Let A and B be the two objects, separated by some obstacle, as a river. From A measure any straight line AC. Fix any point D in the direction AB. Produce AC to F, making $CF = AC$; and produce DC to E, making $CE = CD$. Then find

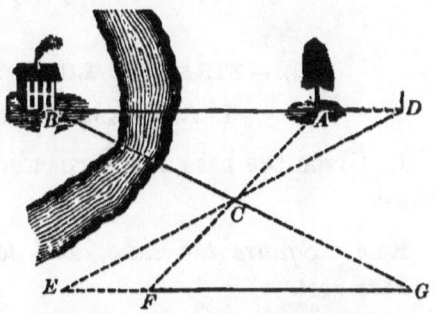

the point G at which the directions of BC and FE intersect; that is, find the point from which C and B appear in one straight line, and E and F appear in another straight line. Then the triangles ACD and CEF are congruent, and therefore ABC and CFG; whence $FG=AB$.

Ww. 106 & 107; (Eu. I. 4 & 26; Cv. I. 76 & 78).

Hence, we find the length of AB by measuring FG.

(γ) *To measure a line when both ends of it are inaccessible.*

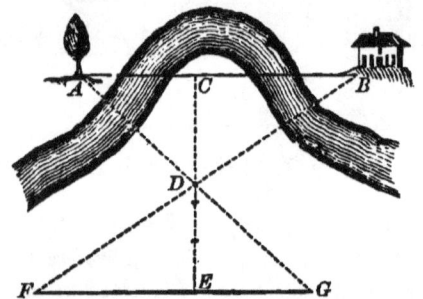

At a point C, in the accessible part of AB, erect a perpendicular CD, and take $DE=CD$. At E make FG perpendicular to DE. Find in FG the point F which falls in the line BD, and the point G in the line AD. $FG=AB$.

Ww. 107; (Eu. I. 26; Cv. I. 78).

(δ) *To measure a line wholly inaccessible.*

If AB is the line, choose a convenient point C from which A and B are both visible, and measure AC and BC by (β); then AB may be measured by (α).

§ (B).—STRAIGHT LINES IN TRIANGLES.

I. RIGHT-ANGLED TRIANGLES.

1. Given the base and perpendicular, to find the hypothenuse.

Rule: *Square the sides, add together, and extract the square root.*

Formula: $a^2 + b^2 = c^2$.

Proof: Ww. 331; (Eu. I. 47; Cv. IV. 25).

THE MEASUREMENT OF LINES.

EXAM. 1. The altitude of a right-angled triangle is 3, the base 4. Find the hypothenuse.

$$a^2 = 3^2 = 9.$$
$$b^2 = 4^2 = 16.$$
$$a^2 + b^2 = 3^2 + 4^2 = 25 = c^2.$$
$$\therefore c = 5. \ Answer.$$

2. Given the hypothenuse and one side, to find the other side.

Rule: *Multiply their sum by their difference, and extract the square root.*

Formula: $c^2 - a^2 = (c+a)(c-a) = b^2.$

EXAM. 2. The hypothenuse of a right-angled triangle is 13, the altitude 12. Find the base.

$$c + a = 25$$
$$c - a = 1$$
$$\therefore c^2 - a^2 = 25 = b^2.$$
$$\therefore b = 5. \ Ans.$$

[For exercises on 1 and 2, see table of right-angled triangles.]

II. OBLIQUE TRIANGLES.

When two lines form an angle, the *projection* of the first on the second is the line between the vertex and the foot of a perpendicular let fall from the extremity of the first on to the second. Thus the projection of AC on BC is CD.

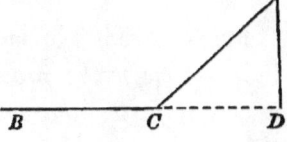

Given two sides and the projection of one on the other, to find the third side:

3. If the angle contained by the given sides be *obtuse.*

Rule: *To the sum of the squares of the given sides add twice the product of the projection and the side on which (when prolonged) it falls; then extract the square root.*

Formula: $a^2 + b^2 + 2bj = c^2.$

Proof: Ww. 336; (Eu. II. 12; Cv. III. 53).

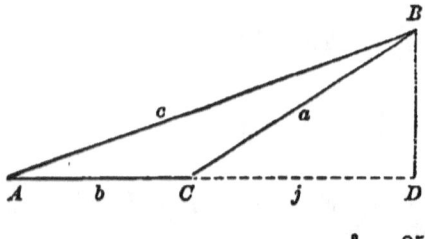

Exam. 3. Given the sides $a=5$, $b=6$, containing an obtuse angle, and given $j=4$, the projection of a on b; find the third side.

$$a^2 = 25$$
$$b^2 = 36$$
$$2bj = 48$$
$$\therefore c^2 = 109.$$

$\therefore c = 10\cdot44+.$ *Ans.*

4. If the angle contained by the given sides is *acute.*

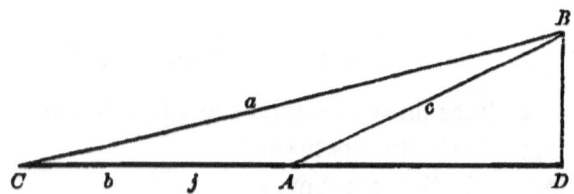

Rule: *From the sum of the squares of the given sides subtract twice the product of the projection and the side on which it falls; the square root of the remainder gives the third side.*

Formula: $a^2 + b^2 - 2bj = c^2.$

Proof: Ww. 335; (Eu. II. 13; Cv. III. 52).

THE MEASUREMENT OF LINES.   9

EXAM. 4. Given the sides $a = 5$, $b = 6$, containing an acute angle, and given $j = 4$, the projection of a on b; find the third side.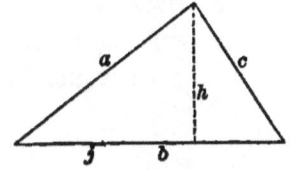
$$a^2 + b^2 = 25 + 36 = 61$$
$$2bj \qquad\quad = 48$$
$$\therefore c^2 = 13$$
$$\therefore c = 3\cdot 60+. \ Ans.$$

5. If two sides and the *perpendicular* let fall on one from the end of the other, are given, the projection can be found by 2, and then the third side by 3 or 4.

If three sides are given, a projection can be found by 3 or 4, and then the perpendicular by 2.

6. Given three sides of a triangle to find its three medials; *i.e.*, the distances from the vertices to the midpoints of the opposite sides.

Rule: *From the sum of the squares of any two sides subtract twice the square of half the base; the square root of half the remainder is the corresponding medial.*

Formula: $a^2 + c^2 - \tfrac{1}{2}b^2 = 2i^2$.

Proof: Ww. 338; (Eu. Appen. 1; Cv. III. 62).

Corollary: Dividing the difference of the squares of two sides by twice the third side, gives the projection on it of its medial.

Formula: $j = \dfrac{a^2 - c^2}{2b}$.

EXAM. 5. Given two sides, $a = 7$, $c = 9$, and the base, $b = 4$. Find the medial.

Here $\quad a^2 + c^2 = 49 + 81 = 130$
$b^2 = 4^2, \ \therefore \tfrac{1}{2}b^2 \qquad\quad = 8$
$\therefore 2i^2 \qquad\qquad\quad = 122$
$\therefore i^2 \qquad\qquad\quad = 61$
$\therefore i = 7\cdot 8. \ Ans.$

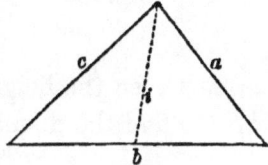

III. STRAIGHT LINES IN SIMILAR FIGURES.

7. Given two straight lines in one figure, and a line corresponding to one of them in a similar figure, to find the line corresponding to the other.

Rule: *The like sides of similar figures are proportional.*

Formula: $b_2 = \dfrac{a_2 b_1}{a_1}$.

Ww. 278; (Eu. VI., Def. I.; Cv. III. 24).

EXAM. 6. The height of an upright stick is 2 meters, and it casts a shadow 3 meters long; the shadow of a flag-staff is 45 meters. Find the height of the staff.

$3 : 2 :: 45 : b_2$.

$\therefore b_2 = \dfrac{90}{3} = 30$ meters. *Ans.*

IV. CHORDS OF A CIRCLE.

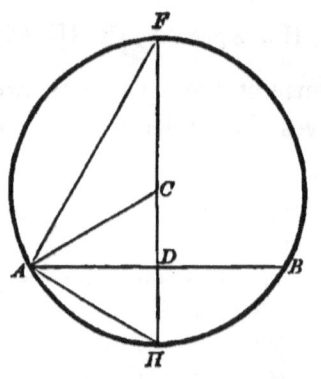

Suppose AB any chord in a circle. Through the center C a diameter perpendicular to AB meets it at its middle point D, and bisects the arc at H. DH is the height of the arc, and AH the chord of half the arc.

8. Given the height of an arc and the chord of half the arc, to find the diameter of the circle.

Rule: *Divide the square of the chord of half the arc by the height of the arc.*

Formula: $d = \dfrac{k_{\frac{1}{2}}^{2}}{h}$.

Proof: HAF is a right angle.

Ww. 204; (Eu. III. 31; Cv. II. 59).

$\therefore HF : HA :: HA : HD$.

Ww. 289; (Eu. VI. 8, Cor.; Cv. III. 44).

EXAM. 7. The height of an arc is 2 centimeters, the chord of half the arc is 6 centimeters. Find the diameter.

$$d = \frac{6 \times 6}{2} = 18. \ Ans.$$

9. Given the chord of an arc and the radius of the circle, to find the height of the arc.

Rule: *From the radius subtract the square root of the difference of the squares of the radius and half the chord.*

Formula: $h = r - \sqrt{r^2 - \tfrac{1}{4}k^2}$.

Proof: $HD = HC - DC$.

$HD = h$, $HC = r$, and $DC^2 = AC^2 - AD^2 = r^2 - (\tfrac{1}{2}AB)^2$.

EXAM. 8. The chord of an arc is 240 millimeters, the radius 125 millimeters. Find the height of the arc.

$DC^2 = r^2 - (\tfrac{1}{2}k)^2 = (r + \tfrac{1}{2}k)(r - \tfrac{1}{2}k) = (125 + 120)(125 - 120)$.

$\therefore DC = \sqrt{r^2 - \tfrac{1}{4}k^2} = \sqrt{245 \times 5} = \sqrt{1225} = 35$.

$\therefore h = r - DC = 125 - 35 = 90$ millimeters
$\hphantom{\therefore h = r - DC = 125 - 35\ } = 9$ centimeters. *Ans.*

10. Given the chord and height of an arc, to find the chord of half the arc.

Rule: *Take the square root of the sum of the squares of the height and half the chord.*

Formula: $k_{\frac{1}{2}} = \sqrt{h^2 + \frac{1}{4}k^2}$.

Proof: $AH^2 = HD^2 + AD^2$.

Ww. 331; (Eu. I. 47; Cv. IV. 25).

Exam. 9. Given the chord $= 48$, the height $= 10$. Find the chord of half the arc.

$$h^2 = 100. \quad (\tfrac{1}{2}k)^2 = (24)^2 = 576.$$

$$\therefore k_{\frac{1}{2}}^2 = 676. \quad \therefore k_{\frac{1}{2}} = 26. \ Ans.$$

If, *instead of the height*, the *radius* is given, substitute in 10 for h its value in terms of r and k from 9, and we have

$$k_{\frac{1}{2}} = \sqrt{(r - \sqrt{r^2 - \tfrac{1}{4}k^2})^2 + \tfrac{1}{4}k^2} = \sqrt{2r^2 - 2r\sqrt{r^2 - \tfrac{1}{4}k^2}}.$$

From this follows:

11. Given the chord of an arc and the radius of the circle, to find the chord of half the arc.

Formula: $k_{\frac{1}{2}} = \sqrt{2r^2 - r\sqrt{4r^2 - k^2}}$.

Exam. 10. Calculate the length of the side of a regular dodecagon inscribed in a circle whose radius is 1 meter; that is, find $k_{\frac{1}{2}}$ when r and k are each 1 meter long, for k is here the side of a regular inscribed hexagon, which always equals the radius.

Ww. 391; (Eu. IV. 15, Cor.; Cv. V. 14).

THE MEASUREMENT OF LINES.

Thus r and k being unity, our formula becomes

$$k_{\frac{1}{2}} = \sqrt{2 - \sqrt{4-1}} = 0.51763809. \quad Ans.$$

EXAM. 11. With unit radius, find the length of one side of a regular inscribed polygon of 24 sides.

$$k_{\frac{1}{2}} = \sqrt{2 - \sqrt{4 - (.51763809)^2}} = 0.26105238.$$

And so on with regular polygons of 48, 96, 192, etc., sides.

12. Given the radius of a circle and the side of a regular inscribed polygon, to find the side of the similar circumscribed polygon.

Formula: $t = \dfrac{2kr}{\sqrt{4r^2 - k^2}}$.

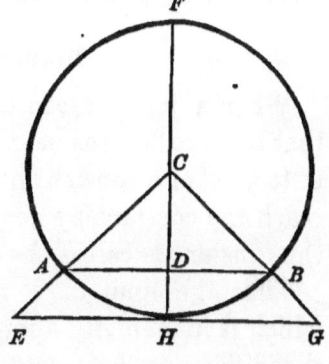

Proof: Suppose AB the given side k. Draw the tangent at the middle point H of the arc AB, and produce it both ways to the points E and G, where it meets the radii CA and CB produced; EG is the side required, t.

In the similar triangles CEH, CAD,

$$CH : CD :: EH : AD :: t : k.$$

$$\therefore t = \frac{kr}{CD}.$$

But $\quad CD^2 = CA^2 - AD^2 = r^2 - (\tfrac{1}{2}k)^2.$

$$\therefore CD = \sqrt{r^2 - \tfrac{1}{4}k^2} = \tfrac{1}{2}\sqrt{4r^2 - k^2}.$$

EXAM. 12. When $r = 1$, find one side of a regular circumscribed dodecagon.

With radius taken as unity, $t = \dfrac{2k}{\sqrt{4-k^2}}$.

From Exam. 10, $k = 0{\cdot}51763809$, and substituting this value, $t_{12} = 0{\cdot}535898$. *Ans.*

In the same way, by substituting $k_{24} = 0{\cdot}26105238$ from Exam. 11, we find $t_{24} = 0{\cdot}263305$, and from $k_{48} = 0{\cdot}13080626$ we get $t_{48} = 0{\cdot}131087$, and so on for 96, 192, 384, etc., sides.

§ (C). **METHOD OF LIMITS.**

V. A variable is a quantity which may have successively an indefinite number of different values.

DEFINITION OF A LIMIT.

When a quantity can be made to vary in such a manner that it approaches as near as we please and continually nearer to a definite constant quantity, but cannot be conceived to reach the constant by any continuation of the process, then the constant is called the *limit* of the variable quantity.

Thus the limit of a variable is the constant quantity which it indefinitely approaches, but never reaches, though the difference between the variable and its limit may become and remain less than any assignable magnitude.

EXAM. 13. The limit of the sum of the series,

$1 + \tfrac{1}{2} + \tfrac{1}{4} + \tfrac{1}{8} + \tfrac{1}{16} + \tfrac{1}{32} + \tfrac{1}{64} +$ etc., is 2.

EXAM. 14. The variable may be likened to a convenient ferry-boat, which will bear us just as close as we choose to the dock,—the constant limit,—but which cannot actually reach or touch it; for, if they touch, both explode into the unknown infinite.

The bridge, the method for passing, in the order of our knowledge, from variables to their limits, is the

THE MEASUREMENT OF LINES.

13.   PRINCIPLE OF LIMITS.

If, while tending toward their respective limits, two variable quantities are always in the same ratio to each other, their limits will be to one another in the same ratio as the variables.*

$A \quad\quad b \quad c \quad\quad\quad B' \quad b' \quad B \quad\quad\quad C' \quad c' \quad C \quad\quad\quad\quad C'''$

Let the lines AB and AC represent the limits of any two variable magnitudes which are always in the same ratio to one another, and let Ab, Ac represent two corresponding values of the variables themselves; then $Ab:Ac::AB:AC$.

If not, then $Ab:Ac::AB:$ some line greater or less than AC. Suppose, in the first place, that $Ab:Ac::AB:AC''$; AC'' being less than AC. By hypothesis, the variable Ac continually approaches AC, and may be made to differ from it by less than any given quantity. Let Ab and Ac, then, continue to increase, always remaining in the same ratio to one another, till Ac differs from AC by less than the quantity $C''C$; or, in other words, till the point c passes the point C'', and reaches some point, as c' between C'' and C, and b reaches the corresponding point b'. Then, since the ratio of the two variables is always the same, we have

By hypothesis, $Ab:Ac::Ab':Ac'$

hence, $Ab:Ac::AB:AC''$;

or, $Ab':Ac'::AB:AC''$

$$AC'' \times Ab' = Ac' \times AB;$$

which is impossible, since each factor of the first member is less than the corresponding factor of the second member.

---

* This principle, and the following demonstration of it, are contained essentially in Eu. XII. 2, though quoted here from Bledsoe.

Hence the supposition that $Ab : Ac :: AB : AC'$, or to any quantity less than AC, is absurd.

Suppose, then, in the second place, that $Ab : Ac :: AB : AC''$, or to some term greater than AC. Now, there is some line, as AB', less than AB, which is to AC as AB is to AC''. If, then, we conceive this ratio to be substituted for that of AB to AC'', we have

$$Ab : Ac :: AB' : AC;$$

which, by a process of reasoning similar to the above, may be shown to be absurd. Hence, if the fourth term of the proportion can be neither greater nor less than AC, it must be equal to AC; or we must have

$$Ab : Ac :: AB : AC.$$ Q. E. D.

Cor.: If two variables are always equal, their limits are equal.

14. *When their sides tend indefinitely toward zero, the perimeter of the polygon inscribed increases, circumscribed decreases, toward the same limit, the length of the curve.*

Proof: Inscribe in a circle any convenient polygon, say, the regular hexagon. Ww. 391; (Eu. IV. 15; Cv. V. 14).

Join the extremities of each side, as AB, to the point of the curve equally distant from them, as H; that is, the point of intersection of the arc and a perpendicular at the middle point of the chord. Thus we get sides of a regular dodecagon. Repeat the process with the sides of the dodecagon, and we have a regular polygon of twenty-four sides. So continuing, the number of sides, always doubling, will increase indefinitely, while the length of a side will tend toward zero.

The length of the inscribed perimeter augments with the number of sides, since we continually replace a side by two which form with it a triangle, and so are together greater.

Ww. 96; (Eu. I. 20; Cv. 5).

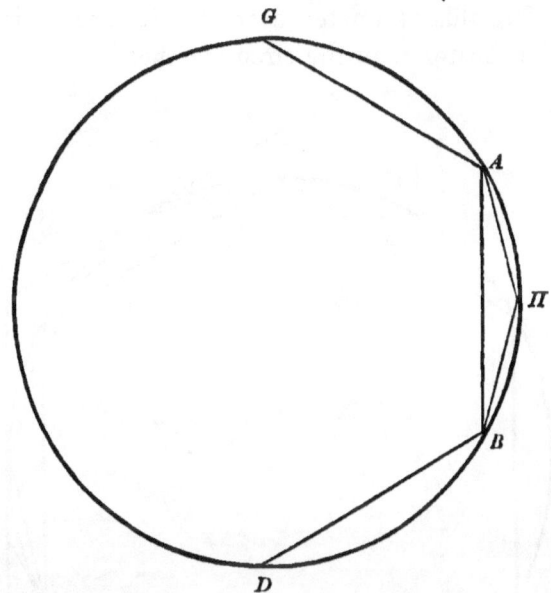

Thus AB of the hexagon is replaced by $A\Pi$ and ΠB in the dodecagon.

But this increasing perimeter can never become as long as the circumference, since it is always made up of chords each of which is shorter than the corresponding arc, by the axiom, "A straight line is the shortest distance between two points." Therefore, this perimeter increases toward a limit which cannot be longer than the circumference.

In doubling the number of sides of a circumscribed polygon, by drawing tangents at the middle points of the arcs, we continually substitute a straight for a broken line; as, TU for $TN + NU$. So this perimeter decreases.

But two tangents from an external point cannot be together shorter than the included arc.

E.g., $\qquad HT + TB >$ arc HB.

Therefore this perimeter decreases toward a limit which cannot be shorter than the circumference.

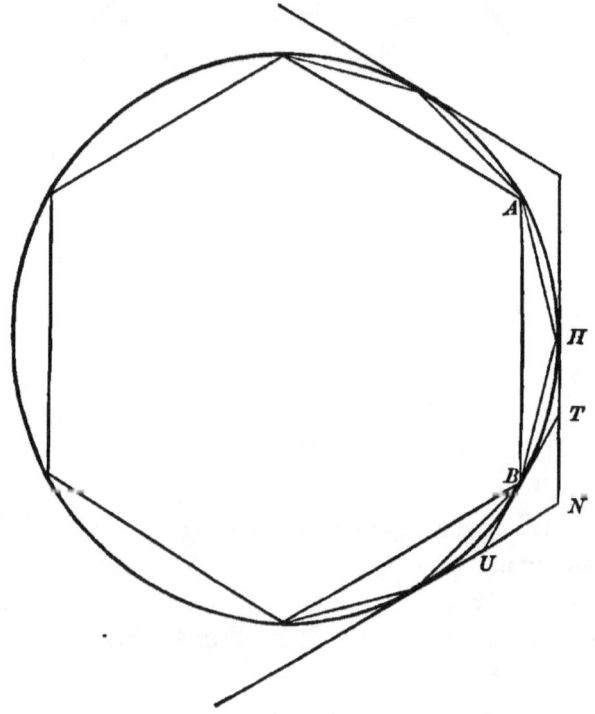

But the limit toward which the circumscribed perimeter decreases is identical with that toward which the corresponding inscribed perimeter increases.

For, in a regular circumscribed polygon of any number of sides, n, the perimeter is n times one of the sides.

$$p_n = nt_n.$$

THE MEASUREMENT OF LINES.

But, from 12,
$$t_n = \frac{2rk_n}{\sqrt{4r^2 - k_n^2}}.$$
$$\therefore p_n = \frac{2nrk_n}{\sqrt{4r^2 - k_n^2}}.$$

But p'_n, the perimeter of the corresponding inscribed polygon, is nk_n.
$$\therefore \frac{p_n}{p'_n} = \frac{2r}{\sqrt{4r^2 - k_n^2}},$$

the ratio of the perimeters.

Cutting the circumference into n equal parts makes each part as small as we please by taking n sufficiently great.

But chords are shorter than their arcs; therefore k_n tends toward the limit zero as n increases.

Thus the limit of $\dfrac{2r}{\sqrt{4r^2 - k_n^2}}$ is $\dfrac{2r}{\sqrt{4r^2}} = 1$. The variables $\dfrac{p_n}{p'_n}$ and $\dfrac{2r}{\sqrt{4r_2 - k_n^2}}$ are always equal. Therefore, by 13, *Cor.*, their limits are equal, and limit of $\dfrac{p_n}{p'_n} = 1$.

$\therefore$ lim. $p_n =$ lim. p'_n.

But we have shown that lim. p'_n cannot be longer than c, and lim. p_n cannot be shorter than c. Therefore, the common limit is c, the length of the curve.

§ (D). THE RECTIFICATION OF THE CIRCLE.

15. *In a circle whose radius is unity, to find the length of the semicircumference.*

From 14, an approximate value of the semicircumference in this circle is given by the semiperimeter of every polygon inscribed or circumscribed, the latter being in excess, and the former in defect of the true value.

In examples 10, 11, and 12, we have already calculated

the length of a side in the regular inscribed and circumscribed polygons of 12, 24, and 48 sides. Continuing the same process, and in each case multiplying the length of one side by half the number of sides, we get the following table of semiperimeters:

n.	$\tfrac{1}{2}nk_n$.	$\tfrac{1}{2}nt_n$.
6	3·000	3·4641016
12	3·1058285	3·2153903
24	3·1326286	3·1596599
48	3·1393502	3·1460862
96	3·1410319	3·1427146
192	3·1414524	3·1418730
384	3·1415576	3·1416627
768	3·1415838	3·1416101
1536	3·1415904	3·1415970
3072	3·1415921	3·1415937
6144	3·1415925	3·1415929
12288	3·1415926	3·1415927
etc.	etc.	etc.

Since the semicircumference, $\tfrac{1}{2}c$, is always longer than $\tfrac{1}{2}nk_n$ and shorter than $\tfrac{1}{2}nt_n$, therefore its value, correct to seven places of decimals, is 3·1415926.

16. *The circumferences of any two circles are to each other in the same ratio as their radii.*

Proof: The perimeters of any two regular polygons of the same number of sides have the same ratio as the radii of their circumscribed circles.

Ww. 374; (Eu. XII. 1, V. 12; Cv. V. 10).

The inscribed regular polygons remaining similar to each other when the number of sides is doubled, their

THE MEASUREMENT OF LINES. 21

perimeters continue to have the same ratio. Hence, by 13, the limits, the circumferences, have the same ratio as their radii.

Cor. 1. Circumferences are to each other as their diameters.

Cor. 2. Since
$$c : c' :: r : r' :: 2r : 2r',$$
$$\therefore \frac{c}{2r} = \frac{c'}{2r'} = \frac{\frac{1}{2}c}{r}.$$

That is, the ratio of any circumference to its diameter is a constant quantity.

This constant, identical with the ratio of any semicircumference to its radius, is denoted by the Greek letter π.

But, in circle with radius 1, semicircumference we have found $3\cdot1415926+$. Therefore, the constant ratio $\pi = 3\cdot1415926+$.

VI. LINES IN ANY CIRCLE.

17.
$$\pi = \frac{c}{d} = \frac{\frac{1}{2}c}{r} = 3\cdot1415926+.$$

Multiplying both sides of this equation by d, gives

18. $$c = d\pi = 2r\pi.$$

$\therefore$ The diameter of a circle being given, to find the circumference.

Rule: *Multiply the diameter by* π.

NOTE. In practice, for π the approximation $3\frac{1}{7}$ or $\frac{22}{7}$ is generally found sufficiently close. A much more accurate value is $\frac{355}{113}$; easily remembered by observing that the denominator and numerator written consecutively, thus, 11 3 | 3 55, present the first three odd numbers each written twice. The value most used is $\pi = 3\cdot1416$.

19. Dividing 18 by π gives

$$d = 2r = \frac{c}{\pi} = c \times \frac{1}{\pi} = c \times 0\cdot3183098+.$$

CHAPTER II.

THE MEASUREMENT OF ANGLES.

§ (E). THE NATURAL UNIT OF ANGLE.

To say "all right angles are equal," assumes that the amount of turning necessary to take a straight line or direction all around into its first position is the same for all points.

Thus the *natural unit* of reference for angular magnitude is one whole revolution, called a *perigon*, and equal to four right angles.

VII. A revolving radius describes equally an angle, a surface, and a curve. Moreover, the perigon, circle, and circumference are each built up of congruent parts; and any pair of angles or sectors have the same ratio as the corresponding arcs.

Ww. 201; (Eu. VI. 33; Cv. II. 51).

Therefore, $\dfrac{\text{any angle}}{\text{perigon}} = \dfrac{\text{its intercepted arc}}{\text{circumference}}$.

That is, if we adopt the whole circumference as the unit of arc;

20. *The numerical measure of an angle at the center of a circle is the same as the numerical measure of its intercepted arc.*

THE MEASUREMENT OF ANGLES.

And this remains true, if, to avoid fractions, we adopt, as practical units of angle and arc, some convenient part of these natural units. The Egyptian astronomers divided the whole circle into 360 equal parts, called degrees; each of these degrees was divided into 60 parts, called minutes; these again into 60 parts, called seconds. These numbers have very convenient factors, being divisible by 1, 2, 3, 4, 5, 6, etc.

Exam. 15.
$$\text{A perigon} = 359° 60' \text{ of angle.}$$
$$\text{A circumference} = 359° 59' 60'' \text{ of arc.}$$

VIII. Hence we say, *An angle at the center is measured by its intercepted arc;* meaning, *An angle at the center is such part of a perigon as its intercepted arc is of the whole circumference.*

§ (F). CIRCULAR MEASURE OF AN ANGLE.

Half a perigon is a flat angle; hence, halving the denominators in VII., and using $\angle$ to mean *angle*, gives

$$\frac{\text{any } \angle}{\text{flat } \angle} = \frac{\text{its intercepted arc}}{\text{semicircumference}}.$$

But, from 18, in every circle, $\tfrac{1}{2}c = r\pi$.
Therefore, dividing denominators by π, gives

$$\frac{\text{any } \angle}{\tfrac{1}{\pi} \cdot \text{flat } \angle} = \frac{\text{length of its arc}}{r}.$$

If, now, we adopt as unit angle that part of a perigon denoted by $\frac{\text{flat } \angle}{\pi}$; that is, the $\angle$ subtended at the center of every circle by an arc equal in length to its radius, and hence named a radian, then, by 20,

21. *The number which expresses any angle in radians also expresses its intercepted arc in terms of the radius.*

So, in terms of whatever arbitrary unit of length the arc and radius may be expressed, if u denote the number of radians in an angle, then, for every ∡,

$$u = \frac{l}{r}.$$

Thus the same angle will be denoted by the same number, whatever be the unit of length employed.

u, or the fraction *arc divided by radius*, is called the circular measure of an ∡.

EXAM. 16. Find the circular measure of *f*.

Here, the arc being a semicircumference, its length $l = r\pi$.

$$\therefore u = \frac{r\pi}{r} = \pi. \text{ Ans.}$$

This is obviously correct, since dividing a flat ∡ by π first gave us our radian.

22. Given the number of degrees in an angle, to find the length of the arc intercepted by it from a given circumference.

Rule: *Multiply the length of the circumference by the number of degrees in the angle, and divide the product by 360.*

Formula: $l = \dfrac{cg°}{360°}.$

Proof: From VII. we have $360° : g° :: c : l$.

NOTE. If the ∡ be given in minutes, the formula becomes $l = \dfrac{cg'}{21600'}$. If in seconds, $l = \dfrac{cg''}{1296000''}$.

THE MEASUREMENT OF ANGLES.

EXAM. 17. How long is the arc of one degree in a circumference of 25,000 miles?
$$l = \frac{2500}{36} = 69{\cdot}4+. \ Ans.$$

23. Given the length of an arc of a given circumference, to find the number of degrees it subtends at the center.

Rule : *Multiply the length of the arc by* 360, *and divide the product by the length of the circumference.*

Formula: $g° = \dfrac{l\,360°}{c}$.

NOTE. To find the number of minutes or seconds:

Formulae: $g' = \dfrac{l\,21600'}{c}$; and $g'' = \dfrac{l\,1296000''}{c}$.

EXAM. 18. Find the number of degrees subtended in any circle by an arc equal to the radius.

Here $g° = \dfrac{l\,360°}{c}$ becomes $\dfrac{r\,360°}{2r\pi} = \dfrac{180°}{\pi}$
$$= 57{\cdot}2957795°+. \ Ans.$$
Hence a radian $= \rho = 57°\,17'\,44{\cdot}8''+ = 206264{\cdot}8''+$.

IX. The arcs used throughout as corresponding to the angles are those intercepted from circles whose center is the angular vertex.

These arcs are said to measure the angles at the center which include them, because these arcs contain their radius as often as the including angle contains the radian.

Using *measured* in this sense, we may state the following Theorems:

24. An angle at the center is measured by the arc intercepted between its sides.
Ww. 202; (Cv. II. 52).

25. An inscribed angle is measured by half its intercepted arc.

Ww. 203; (Eu. III. 20; Cv. II. 57).

26. An angle formed by a tangent and a chord is measured by half the intercepted arc.

Ww. 209; (Eu. III. 32; Cv. II. 62).

27. An angle formed by two chords, intersecting within a circle, is measured by half the sum of the arcs vertically intercepted.

Ww. 208; (Eu. Appen. 2; Cv. II. 64).

28. If two secants, two tangents, or a tangent and a secant intersect without the circle, the angle formed is measured by half the difference of the intercepted arcs.

Ww. 210; (Eu. Appen. 3; Cv. II. 65).

29. Given the measure of an angle in degrees, to find its circular measure.

Rule: *Multiply the number of degrees by π, and divide by* 180.

Formula: $u = \dfrac{\pi g^\circ}{180^\circ} = \dfrac{\pi g'}{10800'} = \dfrac{\pi g''}{648000''}.$

Proof: A flat ∡ is 180°, and its circular measure is π. Hence,

$$\frac{g^\circ}{180^\circ} = \frac{u}{\pi},$$

since each fraction expresses the ratio of any given ∡ to a flat ∡. Therefore,

$$u = \frac{\pi g^\circ}{180^\circ},$$

and also

$$g^\circ = \frac{u \, 180^\circ}{\pi}.$$

THE MEASUREMENT OF ANGLES. 27

This recalls to mind again that *the circular measure of any ∡ is independent of the length of the radius of the circle.*

EXAM. 19. Find the circular measure of ∡ of one degree.
Here $$u = \frac{\pi}{180} = \cdot 0174532925 +. \ Ans.$$

EXAM. 20. Find the circular measure of ∡ of one minute.
Dividing the last answer by 60 gives
$$\cdot 000290888208 +. \ Ans.$$

Of course, this number equally expresses the length of an arc of one minute in parts of the radius, and in the same way we obtain

$$\text{Arc } 1'' = r \times 0 \cdot 00000484813681 +.$$

30. Given the circular measure of an angle, to find its measure in degrees.

Rule: *Multiply the circular measure by 180, and divide by π.*

Formula: $g° = \dfrac{u \, 180°}{\pi}$.

EXAM. 21. Find the number of degrees in ∡ whose circular measure is 10.
Here $$g° = 10 \times \frac{180}{\pi} = 10 \times 57 \cdot 2957795 +$$
$$= 572 \cdot 957795 +. \ Ans.$$

31. Given the angle in degrees and the length of the arc which subtends it, to find the radius.

Rule: *Divide 180 times the length by π times the number of degrees.*

Formula: $r = \dfrac{l \, 180°}{\pi g°}$.

Proof: $\dfrac{g°}{180°} = \dfrac{u}{\pi} = \dfrac{l}{r\pi}.$

Exam. 22. An arc of 6 meters subtends ∡ of 10°, find radius.

Here $\quad r = \dfrac{180}{\pi} \times \dfrac{6}{10} = 0{\cdot}6 \times 57{\cdot}2957795+$
$\quad\quad\quad\quad\quad\quad\quad\quad\quad = 34{\cdot}3775$ meters. *Ans.*

X. One ∡ is called the *complement* of another, when their sum equals a rt. ∡; the *supplement*, when their sum $= f$; the *explement*, when their sum $= \delta$.

Reference Table of Abbreviations for Areas.

When used alone, as abbreviations, *capital* letters *denote* the *area* of the figures; to denote volume, a V is prefixed.

A = annulus.
B = base.
C = cylinder.
$\odot$ = circle.
D = volume of prismatoid.
E = ellipse.
F = frustum.
G = segment.
H = sphere.
I = volume of an irregular polyhedron.
J = parabola.
K = cone.
L = lune.
M = midsection.
N = polygon of n sides.
O = solid ring.
P = prism.
$\square$ = parallelogram.
Q = quadrilateral.
q = square.
R = rectangle.
S = sector.

T = trapezoid.
$\triangle$ = triangle.
U = volume of quader.
V = volume.
W = volume of wedge.
X = volume of tetrahedron.
Y = pyramid.
Z = zone.
$\sum\limits_{v=1}^{v=n} T_v = T_1 + T_2 + T_3 + \cdots + T_n.$
$\mathfrak{t}$ = sum of angles.
$\widehat{\triangle}$ = spherical $\triangle$.
$\widehat{N}$ = spherical N.
$\widehat{Y}$ = volume of spherical Y.
Ω = steregon.
$=$ means equivalent; *i.e.*, equal in size.
$\parallel$ = parallel.
$\perp$ = perpendicular.
$\sim$ = similar.
$\therefore$ = therefore.
∡ = angle.

CHAPTER III.

THE MEASUREMENT OF PLANE AREAS.

§ (G). PLANE RECTILINEAR FIGURES.

XI. The *area* of a surface is its *numerical measure*.

Measuring the area of a surface, whether plane or curved, is determining its *ratio* to a chosen surface called the unit of area.

The chosen unit of area is a square whose side is a unit of length.

EXAM. 23. If the unit of length be a meter, the unit of area will be called a square meter (^{qm}).

If the unit of length be a centimeter, the unit of area will be a square centimeter (^{qcm}).

Square Centimeter

32. To find the area of a rectangle.

Rule : *Multiply the base by the altitude.*

Formula : $R = ab$.

Proof : SPECIAL CASE. When the base and altitude, or length and breadth of the rectangle are *commensurable*.

In this case there is always a line which will divide both base and altitude exactly.

If this line be assumed as linear unit, a and b are integral numbers.

In the rectangle $ABCD$ divide AD into a, and AB into b equal parts. Through the points of division draw lines parallel to the sides of the rectangle. These lines divide the rectangle into a number of squares, each of which is a unit of area. In the bottom row there are b such squares; and, since there are a rows, we have b squares repeated a times, which gives, in all, ab squares.

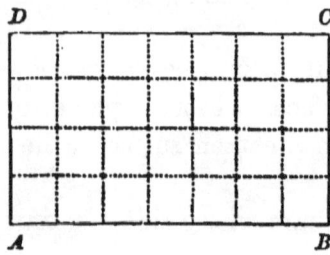

NOTE. The composition of ratios includes numerical multiplication as a particular case. Ordinary multiplication is a growth from addition. The multiplier indicates the number of additions or repetitions. The multiplicand indicates the thing added or repeated. Therefore, it is not a mutual operation, and the product is always in terms of the unit of the multiplicand. The multiplicand may be any aggregate; the multiplier is an aggregate of repetitions. To repeat a thing does not change it in kind, so the result is an aggregate of the same sort exactly as the multiplicand. When the rule says, Multiply the base by the altitude, it means, Multiply the *numerical measure* of the base by the *number measuring* the altitude in terms of the same linear unit. The product is *a number* which we have shown to be the *area* of the rectangle; that is, its numerical measure in terms of the superficial unit.

This is the meaning to be assigned whenever we speak of the product of one line by another.

GENERAL PROOF.

Rectangles, being equiangular parallelograms, have to one another the ratio which is compounded of the ratios of their sides.   Ww. 315; (Eu. VI. 23; Cv. IV. 5).

Let R and R' represent the surfaces or areas of two rectangles. Let a and a' represent their altitudes; b and b' their bases.

THE MEASUREMENT OF PLANE AREAS.

Thus,
$$\frac{R}{R'} = \frac{a}{a'} \times \frac{b}{b'} = \frac{ab}{a'b'}.$$
$$\therefore R = R' \frac{ab}{a'b'}.$$

For the measurement of surfaces, this equation is fundamental. To apply it in practice, we have only to select as a standard some particular unit of area.

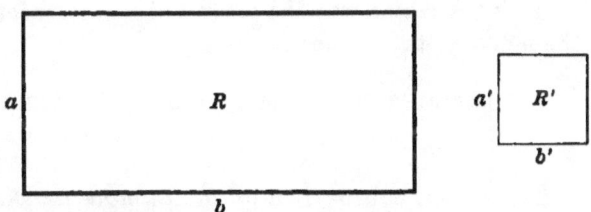

The equation itself points out as best the unit we have already indicated. If we suppose a' and b' to be, each of them, a unit of length, R' becomes this superficial unit, and the equation becomes

$$R = ab.$$

This shows that the number of units of area in any rectangle is that number which is the product of the numbers of units of length in two adjacent sides.

This proof includes every case which can occur, whether the sides of the rectangle be commensurable or incommensurable with the unit of length; that is, whether a and b are integral, fractional, or irrational.

EXAM. 24. Find the area of a ribbon 1 meter long and 1 centimeter wide.

1 meter is 100 centimeters.

∴ 100 square centimeters. *Ans.*

33. To find the area of a square.

Rule: *Take the second power of the number denoting the length of its side.*

Formula: $q = b^2$.

Proof: A square is a rectangle having its length and breadth equal.

NOTE. This is the reason why the product of a number into itself is called the square of that number.

Cor. Given the area of a square, to find the length of a side.

Rule: *Extract the square root of the number denoting the area.*

EXAM. 25. 1 square dekameter, usually called an Ar (ᵃ), contains 100 square meters. Every unit of surface is equivalent to 100 of the next lower denomination, because every unit of length is 10 of the next lower order. Thus a square hektometer is a hektar (ʰᵃ).

EXAM. 26. How many square centimeters in 10 square millimeters?

100 square millimeters is 1 square centimeter.

∴ 10 square millimeters is $\frac{1}{10}$ of 1 square centimeter. *Ans.*

Or, 10 square millimeters make a rectangle 1 centimeter long and 1 millimeter wide.

EXAM. 27. How many square centimeters in 10 millimeters square?

One. *Ans.*

THE MEASUREMENT OF PLANE AREAS. 33

REMARK. *Distinguish carefully between square meters and meters square.*

We say 10 square kilometers (^{qkm}), meaning a surface which would contain 10 others, each a square kilometer; while the expression 5 kilometers square means a square whose sides are each 5 kilometers long, so that the figure contains 25 square kilometers.

EXAM. 28. The area of a square is 1000 square meters. Find its side.
$$\sqrt{1000} = 31\cdot 623 \text{ meters. } Ans.$$

34. To find the area of any parallelogram.

Rule: *Multiply the base by the altitude.*

Formula: $\square = ab$.

Proof: Any parallelogram is equivalent to a rectangle of the same base and altitude.

Ww. 321; (Eu. I. 35; Cv. IV. 10).

Cor. The area of a parallelogram, divided by the base, gives the altitude; and the area, divided by the altitude, gives the base.

EXAM. 29. Find the area of a parallelogram whose base is 1 kilometer, and altitude 1 centimeter.

$$b = 1000 \text{ meters.} \qquad a = \tfrac{1}{100} \text{ meter.}$$

$$\therefore ab = 10 \text{ square meters. } Ans.$$

35. Given one side and the perpendicular upon it from the opposite vertex, to find the area of a triangle.

Rule: *Take half the product of the base into the altitude.*

Formula: $\triangle = \tfrac{1}{2} ab$.

Proof: A triangle is equivalent to half a parallelogram having the same base and altitude.

Ww. 324; (Eu. I. 41; Cv. IV. 13).

Cor. 1. If twice the number expressing the area of a triangle be divided by the number expressing the base, the quotient is the altitude; and *vice versa*.

Cor. 2. Two △'s or □'s, having an equal ∡, are as the products of the sides containing it.

EXAM. 30. One side of a triangle is 35·74 meters, and the perpendicular on it is 6·3 meters. Find the area.

$\frac{1}{2}b = 17\cdot87$ meters.

∴ $\frac{1}{2}ab = 112\cdot581$ square meters. *Ans.*

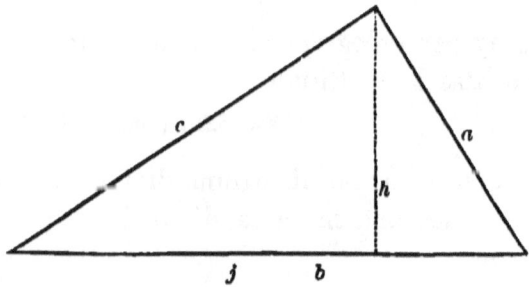

36. Given the three sides of a triangle, to find the area.

Rule: *From half the sum of the three sides subtract each side separately; multiply the half sum and the three remainders together: the square root of the product will be the area.*

Formula: $\triangle = \sqrt{s(s-a)(s-b)(s-c)}$.

Proof:

By 4, $\quad a^2 = b^2 + c^2 - 2bj$,

whence $\quad j = \dfrac{b^2 + c^2 - a^2}{2b}$.

THE MEASUREMENT OF PLANE AREAS.

By 2, $\quad h^2 = c^2 - j^2 = c^2 - \dfrac{(b^2 + c^2 - a^2)^2}{4b^2}$;

whence, $\quad 4b^2h^2 = 4b^2c^2 - (b^2 + c^2 - a^2)^2$.

$\therefore 2bh = \sqrt{4b^2c^2 - (b^2 + c^2 - a^2)^2}$.

$\therefore 2bh = \sqrt{(2bc + b^2 + c^2 - a^2)(2bc - b^2 - c^2 + a^2)}$.

$\therefore 2bh = \sqrt{(a+b+c)(b+c-a)(a+b-c)(a-b+c)}$.

But, by 35, $2bh$ equals four times the area of the triangle.

Cor. To find the area of an equilateral triangle, multiply the square of a side by 0·433+.

EXAM. 31. Find the area of an isosceles triangle whose base is 60 meters and each of the equal sides 50 meters. Here, from last formula in Proof,

$$2bh = b\sqrt{(2a+b)(2a-b)} = 60\sqrt{160 \times 40} = 60\sqrt{1600 \times 4}.$$
$$\therefore 2bh = 60 \times 40 \times 2.$$

$\therefore$ Area $= 1200$ square meters. *Ans.*

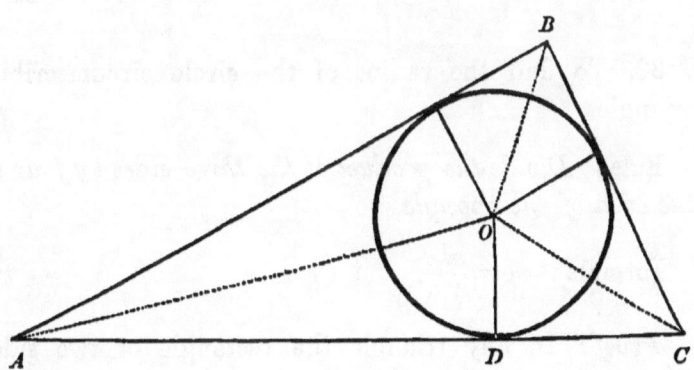

✗ 37. To find the radius of the circle inscribed in a triangle.

Rule: *Divide the area of the triangle by half the sum of its sides.*

Formula: $r = \dfrac{\Delta}{s}$.

Proof.

By 35,
$$\text{area of } BOC = \frac{ar}{2},$$
$$\text{area of } COA = \frac{br}{2},$$
$$\text{area of } AOB = \frac{cr}{2}.$$

∴ by addition, $\triangle = \dfrac{(a+b+c)}{2} \cdot r = sr.$

Cor. The area of any circumscribed polygon is half the product of its perimeter by the radius of the inscribed circle.

Exam. 32. Find radius of circle inscribed in the triangle whose sides are 7, 15, 20.

Here
$$s = 21, \; \therefore \triangle = \sqrt{21 \times 14 \times 6} = \sqrt{3 \cdot 7 \cdot 7 \cdot 2 \cdot 2 \cdot 3}.$$
$$\therefore \triangle = 3 \times 7 \times 2 = 42.$$
$$\therefore r = 2. \; Ans.$$

38. To find the radius of the circle circumscribing a triangle.

Rule: *Divide the product of the three sides by four times the area of the triangle.*

Formula: $\mathfrak{R} = \dfrac{abc}{4\triangle}.$

Proof: In any triangle the rectangle of two sides is equivalent to the rectangle of the diameter of the circumscribed circle by the perpendicular to the base from the vertex. Ww. 300; (Eu. VI. C.; Cv. III. 65).

$$\therefore ac = 2\mathfrak{R}h.$$
$$\therefore \mathfrak{R} = \frac{ac}{2h} = \frac{abc}{2bh}.$$

Cor. The side of an equilateral $\triangle$, $b = \Re \sqrt{3} = 2r\sqrt{3}$.

EXAM. 33. Find radius of circle circumscribing triangle 7, 15, 20.
Here $\qquad abc = 2100, \quad \triangle = 42.$

$$\therefore \Re = \frac{2100}{168} = 12\tfrac{1}{2}. \ Ans.$$

✗ **39.** To find the radius of an escribed circle.

Rule : *Divide the area of the triangle by the difference between half the sum of its sides and the tangent side.*

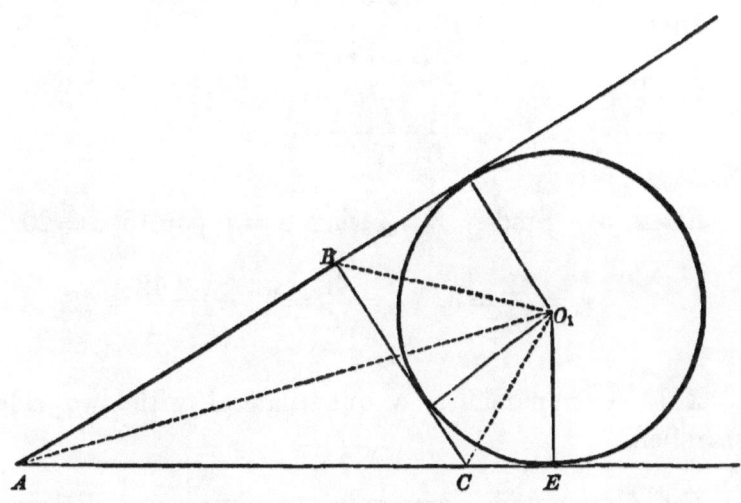

Proof: Let r_1 denote the radius of the escribed circle which touches the side a. The quadrilateral O_1BAC may be divided into the two triangles, O_1AB and O_1AC;

$\therefore$ by 35, $\qquad$ its area $= \dfrac{c}{2}r_1 + \dfrac{b}{2}r_1.$

But the same quadrilateral is composed of the triangles O_1BC and ABC; $\quad \therefore$ its area $= \dfrac{a}{2}r_1 + \triangle.$

Thus,
$$\frac{c}{2}r_1 + \frac{b}{2}r_1 = \frac{a}{2}r_1 + \Delta.$$
$$\therefore \frac{c+b-a}{2}r_1 = \Delta.$$
$$\therefore r_1 = \frac{\Delta}{s-a}.$$

In the same way,
$$r_2 = \frac{\Delta}{s-b}; \quad r_3 = \frac{\Delta}{s-c}.$$

Cor. 1. Since, by 37, $r = \frac{\Delta}{s}$, therefore,

$$r\,r_1 r_2 r_3 = \frac{\Delta^4}{s(s-a)(s-b)(s-c)} = \Delta^2, \qquad \text{by 36.}$$

Thus,
$$\Delta = \sqrt{r\,r_1 r_2 r_3}.$$

Cor. 2.
$$\frac{1}{r_1} + \frac{1}{r_2} + \frac{1}{r_3} = \frac{1}{r}.$$

EXAM. 34. Find r_1, r_2, r_3, when $a = 7$, $b = 15$, $c = 20$.

$$r_1 = \frac{42}{14} = 3, \quad r_2 = \frac{42}{6} = 7, \quad r_3 = \frac{42}{1} = 42. \text{ Ans.}$$

XII. A trapezoid is a quadrilateral with two sides parallel.

Cor. A triangle is a trapezoid one of whose parallel sides has become a point.

40. To find the area of a trapezoid.

Rule: *Multiply the sum of the parallel sides by half their distance apart.*

Formula: $T = x\dfrac{y_1 + y_2}{2}.$

THE MEASUREMENT OF PLANE AREAS.

Proof: Let E be the midpoint of the side AB. Through B and E draw BH and GF parallel to CD. Then $\triangle AEG = \triangle BEF$. Ww. 107; (Eu. I. 26; Cv. I. 78).

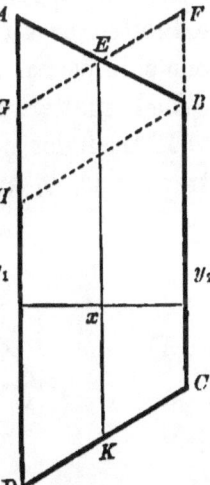

∴ Trapezoid $ABCD = \Box\, GFCD$.
That is, by 34, $T = GD \times x$; where x is the distance of BC from AD. But $DH = BC$, and $HG = AG$.

$$\therefore GD = \frac{AD + BC}{2},$$

and calling AD, y_1, and BC, y_2, we have

$$GD = EK = \frac{y_1 + y_2}{2}.$$

$$\therefore T = \frac{y_1 + y_2}{2} \times x.$$

Cor. The area of a trapezoid equals the distance apart of the parallel sides multiplied by the line joining the midpoints of the non-parallel sides.

Exam. 35. Find the area of a trapezoid whose ‖ sides or bases are 12·34 meters apart, and 56·78 meters and 90· meters long.

```
  56·78 + 90· = 146·78
  12·34 ÷  2 =   6·17
                102746
                14678
                88068
              905·6326 square meters. Ans.
```

✗ XIII. Coördinates of a Point.

The *ordinate* of a point is the perpendicular from it to a fixed base line or *axis*.

The corresponding *abscissa* is the distance from the foot of this ordinate to a fixed point on the axis called the *origin*.

The *coördinates* of any point are its *abscissa* x and its *ordinate* y.

XIV. If to any convenient axis ordinates be dropped from the angular points of any polygon, the *polygon* is exhibited as an algebraic *sum of trapezoids*, each having one side perpendicular to the two parallel sides, and hence called right trapezoids.

If triangles occur, as $1D2$, $6E5$, they are considered trapezoids, y_1 and y_5 being zero.

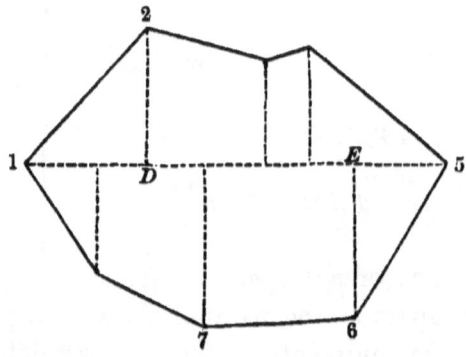

41. To find the sum of any series of right trapezoids.

Rule: *Multiply the distance of* EACH *intermediate ordinate from the first by the difference between its two adjacent ordinates, always subtracting the one following from the one preceding in order along the broken line. Also multiply distance of last ordinate from first by the sum of last two ordinates. Halve the sum of these products.*

Formula: $\sum_{v=1}^{v=n} T_v = \frac{1}{2}[(x_2-x_1)(y_1-y_3)+(x_3-x_1)(y_2-y_4)+\ldots$
$+(x_n-x_1)(y_{n-1}-y_{n+1})+(x_{n+1}-x_1)(y_n+y_{n+1})]$.

Proof: With O as origin, the area of the first trapezoid, by 40, is $(x_2-x_1)\frac{y_1+y_2}{2}$, and of the second is $(x_3-x_2)\frac{y_2+y_3}{2}$.

THE MEASUREMENT OF PLANE AREAS.

Adding the two, we have

$$T_1 + T_2 = \tfrac{1}{2}[(x_2 - x_1)(y_1 + y_2) + (x_3 - x_2)(y_2 + y_3)].$$

Performing the indicated multiplications, $x_2 y_2$ is cancelled by $-x_2 y_2$, and

$$T_1 + T_2 = \tfrac{1}{2}[x_2 y_1 - x_1 y_1 - x_1 y_2 + x_3 y_2 + x_3 y_3 - x_2 y_3].$$
$$\therefore T_1 + T_2 = \tfrac{1}{2}[(x_2 - x_1)(y_1 - y_3) + (x_3 - x_1)(y_2 + y_3)],$$

since here $x_1 y_3$ is balanced by $-x_1 y_3$.

Thus we have proved our rule for a pair of trapezoids. Taking three, we get, by 40,

$$T_3 = (x_4 - x_3)\frac{y_3 + y_4}{2}.$$
$$\therefore T_1 + T_2 + T_3 = \tfrac{1}{2}[(x_2 - x_1)(y_1 - y_3) + (x_3 - x_1)(y_2 + y_3)] + \tfrac{1}{2}[(x_4 - x_3)(y_3 + y_4)].$$

As before, replacing the balancing terms $x_3 y_3 - x_3 y_3$ by $x_1 y_4 - x_1 y_4$, this becomes

$$T_1 + T_2 + T_3 = \tfrac{1}{2}[(x_2 - x_1)(y_1 - y_3) + (x_3 - x_1)(y_2 - y_4) + (x_4 - x_1)(y_3 + y_4)].$$

This proves the rule for three trapezoids; and a generalization of this process proves that if the rule is true of a series of n trapezoids, it is true of $n+1$.

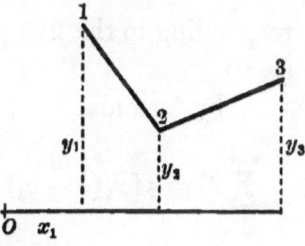

For, by 40, the area of the $(n+1)$th trapezoid

$$T_{n+1} = (x_{n+2} - x_{n+1})\frac{y_{n+1} + y_{n+2}}{2},$$

and, adding this to the first n trapezoids, as given by formula, therefore

$$\sum_{v=1}^{v=n+1} T_v = \tfrac{1}{2}[(x_2 - x_1)(y_1 - y_3) + \cdots \\ + (x_{n+1} - x_1)(y_n + y_{n+1})] + \tfrac{1}{2}[(x_{n+2} - x_{n+1})(y_{n+1} + y_{n+2})].$$

Replacing $x_{n+1}y_{n+1} - x_{n+1}y_{n+1}$ by the balancing terms $x_1 y_{n+2} - x_1 y_{n+2}$, this becomes

$$\sum_{v=1}^{v=n+1} T_v = \tfrac{1}{2}[(x_2 - x_1)(y_1 - y_2) + \cdots \\ + (x_{n+1} - x_1)(y_n - y_{n+2}) + (x_{n+2} - x_1)(y_{n+1} + y_{n+2})].$$

The same method proves that if the formula applies to n trapezoids, it must apply to $n-1$. Therefore, the rule is true for any and every series whatsoever of right trapezoids.

EXAM. 36. Find the right portion of a railroad cross-section whose surface line breaks twice, at the points (x_2, y_2), (x_3, y_3), to the right of center line; (origin being on grade in midpoint of roadbed).

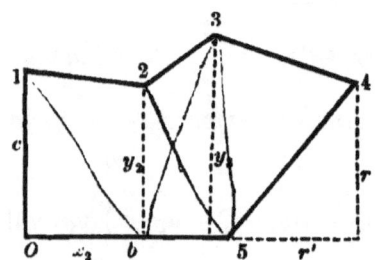

This asks us to find the sum of right trapezoids corresponding to the five points $(0, c)$, (x_2, y_2), (x_3, y_3), $(b + r', r)$, $(b, 0)$.

∴ by formula,

$$\sum_{v=1}^{v=4} T_v = \tfrac{1}{2}[x_2(c - y_3) + x_3(y_2 - r) + (b + r')y_3 + br]. \quad Ans.$$

In the same way, the portion to the left of center line, whether without breaks, or with any number of breaks, is given by our formula, which thus enables us at once to calculate all railroad cross-sections, whether regular or irregular.

EXAM. 37. To find the area of a triangle in terms of the coördinates of its angular points.

THE MEASUREMENT OF PLANE AREAS.

Here are three trapezoids, and consequently, four points; but A is both 1 and 4, so
$$x_4 - x_1 = 0,$$
and the formula becomes

$$\triangle = \sum_{v=1}^{v=3} T_v = \tfrac{1}{2}[(x_2 - x_1)(y_1 - y_3) + (x_3 - x_1)(y_2 - y_1)].$$

$$\therefore \triangle = -\tfrac{1}{2}[x_1 y_2 - x_2 y_1 + x_2 y_3 - x_3 y_2 + x_3 y_1 - x_1 y_3]. \quad Ans.$$

Notice the symmetry of this answer.

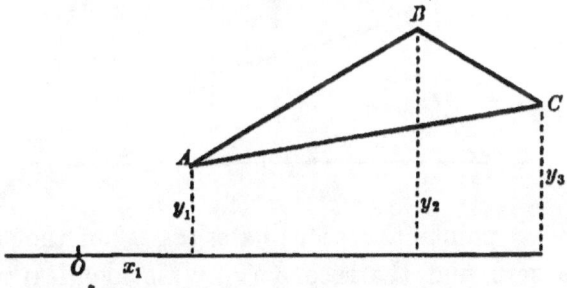

For practical computation this is written

$$2\triangle = x_1(y_3 - y_2) + x_2(y_1 - y_3) + x_3(y_2 - y_1),$$
or $\quad 2\triangle = y_1(x_2 - x_3) + y_2(x_3 - x_1) + y_3(x_1 - x_2).$

To insure accuracy, reckon the area by each.

✗ 42. To find the area of any polygon.

Rule: *Take half the sum of the products of the abscissa of each vertex by the difference between the ordinates of the two adjacent vertices; always making the subtraction in the same direction around the polygon.*

Formula for a polygon of n sides:

$$N = \tfrac{1}{2}[x_1(y_n - y_2) + x_2(y_1 - y_3) + x_3(y_2 - y_4) + \cdots + x_n(y_{n-1} - y_1)].$$

Proof: This is only that special case of 41, where the broken line, being the perimeter of a polygon, ends where it began.

Join the vertex 1 of the polygon 1, 2, 3, 4, ..., n, 1, with the origin O. Then the area enclosed by the perimeter is the same, whether we consider it as starting and stopping at 1 or at O. But, under the latter supposition, though we

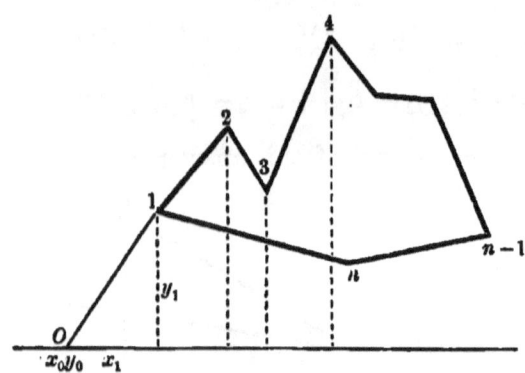

have $n+3$ points, the coördinates (x_0, y_0) of the first and last are zero, and the second (x_1, y_1) is identical with the point next to last; so that formula 41 becomes

$$\tfrac{1}{2}[x_1(0-y_2) + x_2(y_1-y_3) + x_3(y_2-y_4) + \cdots + x_n(y_{n-1}-y_1) + x_1(y_n-0)].$$

NOTE. No mention need be made of minus trapezoids, since the rule automatically gives to those formed by the broken line while going forward, the opposite sign to those formed while going backward.

Our expression for the area of any rectilinear figure is the difference between a set of positive and an equal number of negative terms. If this expression is negative when the angular points are taken in the order followed by the hands of a watch, then it is necessarily positive when they are taken in the contrary sense, for this changes the order in every pair of ordinates in the formula.

Observe that each term is of the form xy, and that there is a pair of these terms, with the minus sign between them, for each vertex of the figure. Thus, for the vertex $(x_m y_m)$, we have the pair $x_m(y_{m-1} - y_{m+1})$; or, pairing those terms which have the same pair of suffixes, for every vertex m, we have $(x_{m+1}y_m - x_m y_{m+1})$. Hence, for twice the area write down the pair $xy - xy$ for each vertex, and add symmetrically the suffixes,

$$1, 2 \quad 2, 1; \quad 2, 3 \quad 3, 2; \quad 3, 4 \quad 4, 3; \quad \ldots\ldots \quad n, 1 \quad 1, n.$$

Thus, for every quadrilateral,

$$2Q = x_1 y_2 - x_2 y_1 + x_2 y_3 - x_3 y_2 + x_3 y_4 - x_4 y_3 + x_4 y_1 - x_1 y_4.$$

But, if any point of perimeter be to the left of origin, or if, to shorten the ordinates, the axis be drawn across the figure, then one or more of the coördinates will be essentially negative. Thus, if in a quadrilateral, we take for axis a diagonal, then

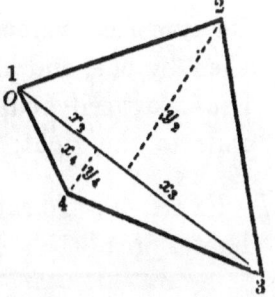

$$x_1 = 0, \quad y_1 = 0, \quad y_3 = 0,$$
and $$2Q = -x_3 y_2 + x_3 y_4.$$

Here, y_4 being essentially negative, the two terms have the same sign, and give the ordinary rule:

43. To find the area of any quadrilateral.

Rule: *Multiply half the diagonal by the sum of the perpendiculars upon it from the opposite angles.*

EXAM. 38. The two diagonals of a quadrilateral measure 1·492 and 37·53 meters respectively, and are $\perp$ to one another. Find the area.

By 43,
$$\text{Area} = \frac{1\cdot 492 \times 37\cdot 53}{2} = \frac{55\cdot 99476}{2}$$
$$= 27\cdot 99738 \text{ square meters. } Ans.$$

46   MENSURATION.

✗ EXAM. 39. Find the area of the polygon 1234567891, the coördinates of whose angular points are (0, 90), (30, 140), (110, 130), (80, 90), (84, 80), (130, 40), (90, 20), (40, 0), (35, 70).
By 42,

$2N = 30 \times -40 + 110 \times 50 + 80 \times 50 + 84 \times 50 + 130 \times 60 + 90 \times 40$
$+ 40 \times -50 + 35 \times -90 = 18750.$

$\therefore$ area $= 9375$. *Ans.*

REMARK. The result of any calculation by coördinates may be verified by a simple change of origin. If the origin is moved to the right through a unit of distance, then the numerical values of all positive abscissae will be diminished by one, and all negative abscissae increased by one. Thus, to verify our last answer, move the origin thirty units to the right, and the question becomes

✗ EXAM. 40. To calculate the area of polygon whose coördinates are:

	x	y
1	−30	90
2	0	140
3	80	130
4	50	90
5	54	80
6	100	40
7	60	20
8	10	0
9	5	70

By 42, twice the area equals the sum of

$-30 \times -70 = 2100$
$80 \times 50 = 4000$
$50 \times 50 = 2500$
$54 \times 50 = 2700$
$100 \times 60 = 6000$
$60 \times 40 = \underline{2400}$
19700
10×-50 and $5 \times -90 = \underline{-950}$
18750

$\therefore$ area $= 9375$, as before.

EXAM. 41. From the data of Exam. 40 construct the figure.

Choose a convenient axis and origin, noticing that the

THE MEASUREMENT OF PLANE AREAS. 47

polygon will lie wholly above the axis, since there are no minus ordinates. Then, to find first vertex, measure off on the axis 30 units to the left of origin, and at the point thus determined, erect a perpendicular 90 units in length. Its

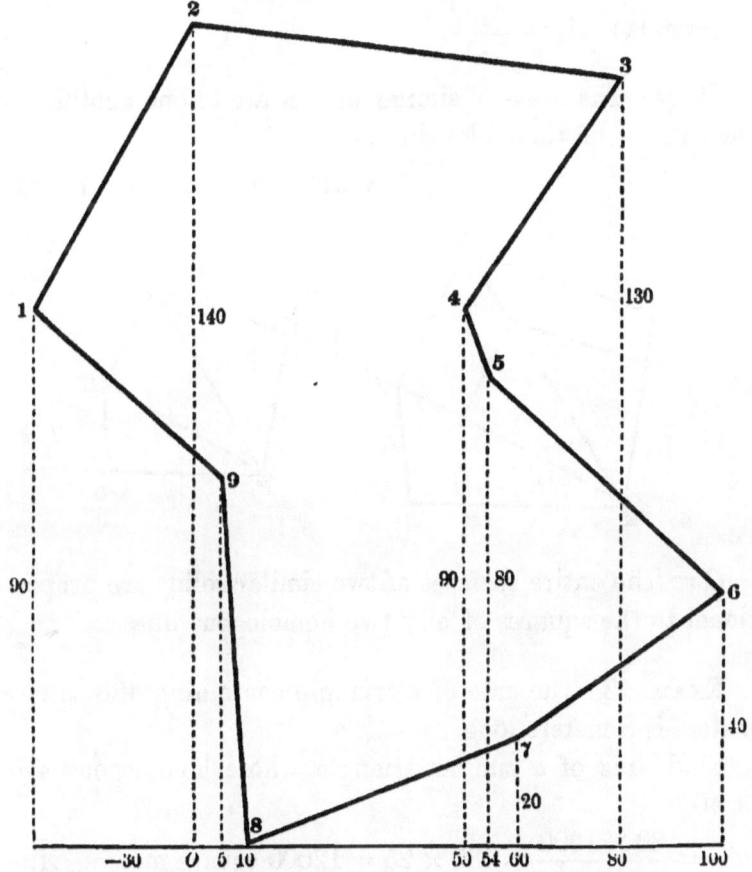

extremity will be the angular point numbered 1. The extremity of a perpendicular at origin 140 units long gives vertex 2, and an ordinate 130 long from a point on axis 80 units to the right of origin gives 3. When all the angular points have been thus determined, join them by straight lines in their order of succession.

44. Given the area and one side of a figure, and the corresponding side of a similar figure, to find its area.

Rule: *Multiply the given area by the squared ratio of the sides.*

Formula: $A_1 = \dfrac{A_2 a_1^2}{a_2^2}$.

Proof: The areas of similar figures are to one another as the squares of their like sides.

Ww. 343; (Eu. VI. 20; Cv. IV. 23).

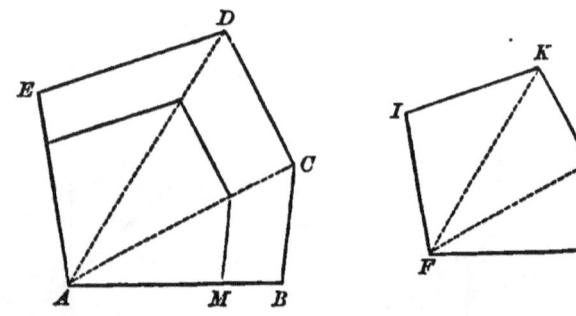

Cor. The entire surfaces of two similar solids are proportional to the squares of any two homologous lines.

Exam. 42. The side of a triangle containing 480 square meters is 8 meters long.

Find area of a similar triangle whose homologous side is 40.

$\triangle = \dfrac{480 \times 1600}{64} = 480 \times 25 = 12000$ square meters. *Ans.*

XV. Magnitudes which can be made to coincide are *congruent*.

Magnitudes which agree in size, but not in shape, are *equivalent*.

THE MEASUREMENT OF PLANE AREAS. 49

XVI. A regular polygon is both equilateral and equiangular.

The bisectors of any two angles of a regular polygon intersect in a point equidistant from all the angular points of the polygon, and hence also equidistant from all the sides, and at once the center of an inscribed and a circumscribed circle.

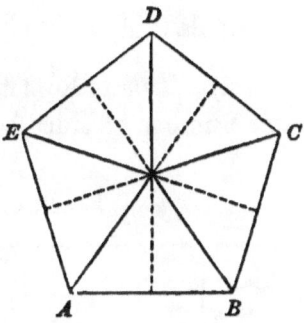

Joining this center to every angle of the polygon cuts it up into congruent isosceles triangles. Hence the area of the regular polygon is the area of any one of these triangles multiplied by the number of sides of the polygon.

45. To find the area of a regular polygon.

Rule: *Multiply together one side, the perpendicular from the center, and half the number of sides.*

Or, in other words:

Take half the product of perimeter by apothem.

Formula: $N = \dfrac{aln}{2} = \dfrac{ap}{2}$.

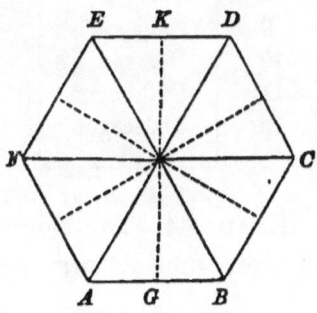

EXAM. 43. The side of a regular hexagon is 98 centimeters, and its apothem 84·87 centimeters; find its area.

$$\text{Area} = 3 \times 98 \times 84\cdot 87$$
$$= 24951\cdot 78 \text{ square centimeters. } Ans.$$

MENSURATION.

46. By the aid of a table of polygons, to find the area of any regular polygon.

Rule: *Multiply the square of one of the sides of the polygon by the area of a similar polygon whose side is unity.*

Formula: $N_i = l_n^2 N_1$.

Proof: This follows from 44, all regular polygons of the same number of sides being similar.

TABLE OF REGULAR POLYGONS.

Number of Sides.	Name.	Area when Side = 1.	Number of Sides.	Area in Terms of Square on Side.
3	Triangle	0·4330127	15	17·642363
4	Square	1·0000000	16	20·109358
5	Pentagon	1·7204774	20	31·568757
6	Hexagon	2·5980762	24	45·574525
7	Heptagon	3·6339124	25	49·473844
8	Octagon	4·8284271	30	71·357734
9	Nonagon	6·1818242	32	81·225360
10	Decagon	7·6942088	40	127·062024
11	Undecagon	9·3656399	48	138·084630
12	Dodecagon	11·1961524		

EXAM. 44. The side of a regular hexagon is 98 centimeters; find its area.

$$\text{Area} = 98 \times 98 \times 2 \cdot 5980762$$
$$= 24951 \cdot 78 \text{ square centimeters. } Ans.$$

EXAM. 45. If the side of a regular decagon is 0·6 meters, its area is

$$0 \cdot 6 \times 0 \cdot 6 \times 7 \cdot 6942088 = 2 \cdot 76991524 \text{ square meters. } Ans.$$

THE MEASUREMENT OF PLANE AREAS. 51

§ (H). AREAS OF PLANE CURVILINEAR FIGURES.

47. To find the area of a circle.

Rule: *Multiply its squared radius by π.*

Formula: $\odot = r^2\pi$.

Proof: If a regular polygon be circumscribed about the circle, its area, by 45, is

$$N = \tfrac{1}{2} r p_n;$$

and, by 14, as n increases, p_n decreases toward c as limit, and N toward $\odot$. But the variables N and p_n are always in the constant ratio $\tfrac{1}{2} r$; therefore, by 13, their limits are in the same ratio, and we have

$$\odot = \tfrac{1}{2} r c.$$

By 18, $\qquad c = 2r\pi.$

Therefore, $\qquad \odot = r^2\pi.$

EXAM. 46. Find the area of a circle whose diameter is 7·5 meters.

Here $\qquad r^2 = 14\cdot0625.$

$\therefore r^2\pi = 44\cdot178+$ square meters. *Ans.*

48. To find the area of a sector.

Rule: *Multiply the length of the arc by half the radius.*

Formula: $S = \tfrac{1}{2} lr = \tfrac{1}{2} u r^2.$

Proof:

Ww. 382; (Cv. V. 44);

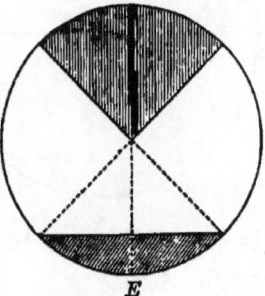

or, as follows: By Eu. VI. 33,

$$S : \odot :: l : c :: u : 2\pi.$$

$$\therefore S = \frac{\odot l}{c} = \frac{\odot u}{2\pi}.$$

$$\therefore S = \frac{\tfrac{1}{2} rcl}{c} = \frac{r^2 \pi u}{2\pi}.$$

$$\therefore S = \tfrac{1}{2} lr = \tfrac{1}{2} r^2 u.$$

EXAM. 47. Find the area of a sector whose arc is 99·58 meters long, and radius 86·34 meters.

$$99{\cdot}58 \times 43{\cdot}17 = 4298{\cdot}8686 \text{ square meters. } Ans.$$

EXAM. 48. Find the area of a sector whose radius is 28 centimeters, and which contains an angle of 50° 36′.

Here, by 29,
$$u = 0{\cdot}883+$$
$$r^2 = 28^2 = 784$$
$$3532$$
$$7064$$
$$6181$$
$$2\overline{)692{\cdot}272}$$
$$\therefore S = 346{\cdot}136 \text{ square centimeters. } Ans.$$

49. To find the area of a segment less than a semicircle.

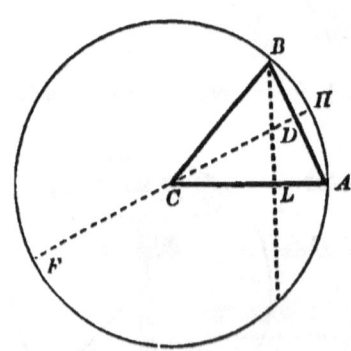

Rule: *From the sector having the same arc as the segment, subtract the triangle formed by the chord and the two radii from its extremities.*

Formula:
$$G = \frac{h^2(l+k) + \tfrac{1}{3} k^2(l-k)}{4h}.$$

Proof: The segment AHB is the difference between the sector $AHBC$ and the triangle ABC.

By 48, $AHBC = \frac{1}{2}lr$.
By 35, $ABC = \frac{1}{2}AB \times CD = \frac{1}{2}k(r-h)$.
$\therefore G = S - \triangle = \frac{1}{2}lr - \frac{1}{2}k(r-h)$.
$\therefore 2G = lr - kr + kh$.

But
$$HD \times DF = AD^2.$$
Ww. 307; (Eu. VI. 13; Cv. III. 47).

$\therefore h(2r-h) = \frac{1}{4}k^2$.
$\therefore r = \dfrac{\frac{1}{4}k^2 + h^2}{2h}$.

Substituting this value of r in the expression for $2G$, we obtain

$$2G = (l-k)\frac{\frac{1}{4}k^2 + h^2}{2h} + kh = \frac{\frac{1}{4}k^2(l-k) + h^2l - h^2k + 2h^2k}{2h}.$$

$$\therefore G = \frac{\frac{1}{4}k^2(l-k) + h^2(l+k)}{4h}.$$

Cor. The area of a segment of a circle is equal to half the product of its radius and the excess of its arc over half the chord of double that arc. For

sector $AHBC = \frac{1}{2}lr$,
and $\triangle ABC = \frac{1}{2}r \times BL$.
$\therefore$ segment $AHB = \frac{1}{2}r(l - BL)$.

Approximate Rule for Segment: *Take two-thirds the product of its chord and height.*

Approximate Formula: $G = \frac{2}{3}hk$.

EXAM. 49. If the chord of a segment is $r \times 0.959851+$, and its height is $r \times 0.122417+$, then an approximation to its area is

$\frac{2}{3}r^2 \times 0.959851+ \times 0.122417+ = \frac{2}{3}r^2 \times 0.117502+ = r^2 \times 0.0783+$.

But if, also, we can measure the arc, and here find it equal to radius, then

$$G = \frac{r^2(0.122417+)^2(r+r\times 0.959851+) + \tfrac{1}{3}r^2(0.959851+)^2(r-r\times 0.959851+)}{4r \times 0.122417+}$$

$\therefore G = \tfrac{1}{4}r^2 \times 0.122417+ + \tfrac{1}{4}r^2 \times 0.117502+ + \tfrac{1}{15}r^2 \times 7.52603+ - \tfrac{1}{15}r^2 \times 7.22387-$
$\therefore G = \tfrac{1}{4}r^2 \times 0.239919+ + \tfrac{1}{15}r^2 \times 0.30216+$
$\therefore G = \tfrac{1}{4}r^2 (0.239919+ + 0.07554+)$
$\therefore G = \tfrac{1}{4}r^2 \times 0.315459+.$

$\therefore G = r^2 \times 0.07886+.$ *Ans.*

Proceeding directly by Rule 49, instead of Formula 49, we here get

$S = \tfrac{1}{2}r^2$,

and $\quad \Delta = \tfrac{1}{2}(r \times 0.959851+)(r - r \times 0.122417+)$
$\therefore \Delta = r \times 0.4799255+ \times r \times 0.877583-$
$\therefore \Delta = r^2 \times 0.421174$
$\therefore G = S - \Delta = r^2 \times 0.07883.$

Since, in this example, arc $= r$, $\therefore G$ is, in any $\odot$, the segment whose ∡ is ρ.

50. A *circular zone* is that part of a circle included between two parallel chords, and may be found by taking the segment on the shorter chord out of that on the longer.

51. A *crescent* is the figure included between the corresponding arcs of two intersecting circles, and is the difference between two segments having a common chord, and on the same side of it.

52. To find the area of an annulus; that is, the figure included between two concentric circumferences.

Rule: *Multiply the sum of the two radii by their difference, and the product by π.*

Formula: $A = (r_1 + r_2)(r_1 - r_2)\pi$.

THE MEASUREMENT OF PLANE AREAS. 55

Proof: By 47, the area of the outer circle is $r_1^2\pi$, and of the inner circle $r_2^2\pi$. Therefore, their difference, the annulus, is $(r_1^2 - r_2^2)\pi$.

Cor. The area of the annular figure will be the same whether the circles are concentric or not, provided one circle is entirely within the other. If the two circles intersect, they form two crescents, one on each side of the common chord, and the difference of the two crescents will always be equal to the annulus formed by the same circles.

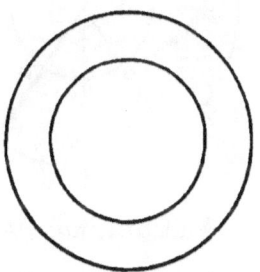

EXAM. 50. The radii of two concentric circles are 39 meters and 11·3 meters. Find the area of the ring between their circumferences.
Here
$$A = 50.3 \times 27.7 \times \pi = 4377.2+ \text{ square meters. } Ans.$$

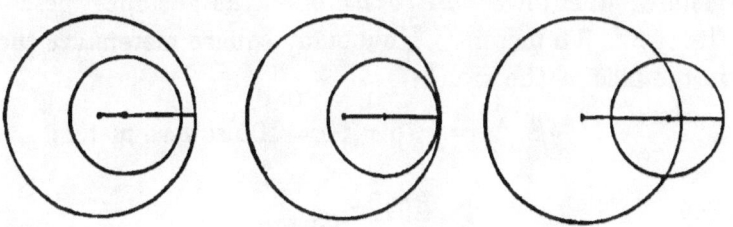

✗ 53. To find the area of a sector of an annulus.

Rule: *Multiply the sum of the bounding arcs by half the distance between them.*

Formula: S. A. $= \frac{1}{2} h(l_1 + l_2)$.

Proof: The sectorial area $ABED$ is the difference between the sector ABC and the sector CDE.

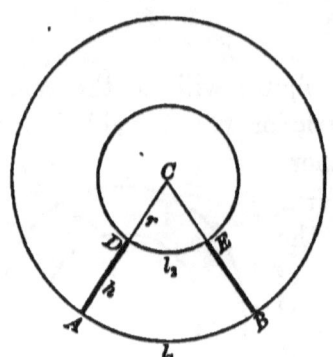

∴ by 48,

$$\text{S. A.} = \tfrac{1}{2}(r+h)l_1 - \tfrac{1}{2}rl_2$$
$$= \tfrac{1}{2}(hl_1 + l_1r - l_2r).$$

Now, since l_1 and l_2 are arcs subtending the same angle at C, ∴ by IX,

$$\frac{l_1}{r+h} = \frac{l_2}{r}.$$
$$\therefore hl_2 = l_1r - l_2r.$$

Substituting, we have

$$\text{S. A.} = \tfrac{1}{2}(hl_1 + hl_2) = \tfrac{1}{2}h(l_1 + l_2).$$

Cor. By comparison with 40, we see an annular sector is equivalent to a trapezoid whose parallel sides equal the arcs, and are at the same distance from one another.

EXAM. 51. The upper arc of a circular arch is 35·25 meters; the lower, 24·75 meters; the distance between the two is 3·5 meters. How many square meters are there in the face of the arch?

Here
$$\text{S. A.} = 1\text{·}75 \times 60 = 105 \text{ square meters. } Ans.$$

✗ XVII. Conics.

If a straight line and a point be given in position in a plane, and if a point move in the plane in such a manner that its distance from the given point always bears the same ratio to its distance from the given line, the curve traced out by the moving point is called a *conic*.

The fixed point is called the *focus*, and the fixed line the *directrix of the conic.*

THE MEASUREMENT OF PLANE AREAS. 57

When the ratio is one of equality, the curve is called a *parabola*.

54. To find the area of a parabolic segment; that is, the area between any chord of a parabola and the part of the curve intercepted.

Rule: *Take two-thirds the product of the chord by the height of the segment.*

Formula: $J = \frac{2}{3} hk$.

Proof: A parabolic segment is two-thirds of the triangle made by the chord and the tangents at its extremities.

If AB, AC, be two tangents to a parabola, to prove that the area between the curve and the chord BC is two-thirds of the triangle ABC.

Parallel to BC draw a tangent DPE. Join A to the point of contact P, and produce AP to cut the chord BC at N.

By a property of the parabola, deducible from its definition,
$$AP = PN.$$
$$\therefore BC = 2 \cdot DE.$$

Ww. 276 & 279; (Eu. VI. 2 & 4; Cv. III. 15 & 25).

$\therefore$ by 35, $\quad\quad \triangle BPC = 2 \cdot ADE$.

This leaves for consideration the two small triangles PDB, PEC, each made by a chord and two tangents. With each proceed exactly as with the original triangle: *e.g.*, draw the tangent FQG parallel to PB; join DQ, and produce it to M; then

$$DQ = QM.$$
$$\therefore PB = 2 \cdot FG.$$
$$\therefore \triangle PQB = 2 \cdot FDG.$$

This leaves four little tangential triangles, like *PFQ*. In each of these draw a tangent parallel to the chord, etc., and let this process be continued indefinitely.

Then the sum of the triangles taken away within the parabola is double the sum of the triangles cut off without

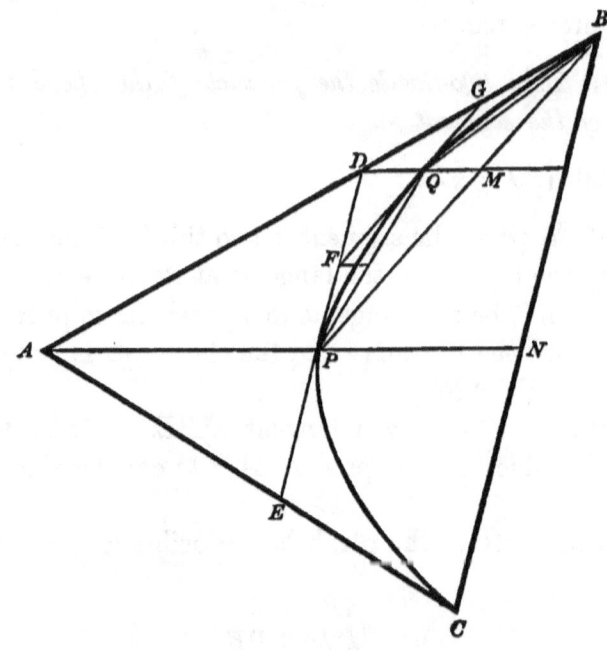

it. But the sum of the interior triangles approaches, as its limit, the parabolic segment. For the triangle *BPC*, since it is half of *ABC*, is greater than half the parabolic area *BQPC*, and so successively with the smaller interior triangles. Therefore, the difference between the parabolic segment and the sum of these triangles can be made less than any assignable quantity.

Ww. 198; (Eu. XII., Lemma; Cv. V. 28).

Therefore, the constant segment is, by definition V., the limit of this variable sum.

THE MEASUREMENT OF PLANE AREAS. 59

Again, each outer triangle cut off is greater than half the area between the curve and the two tangents; *e. g.*, ADE, being half the quadrilateral $ABPC$, is more than half the area $ABQPC$. Therefore, the limit of the sum of the outer triangles is the area between the curve and the two tangents AB, AC. But these two variable sums are always to each other in the constant ratio of 2 to 1. Therefore, by 13, their limits are to each other in the same ratio, and the parabolic segment is two-thirds its tangential triangle.

But the altitude of this triangle is twice the height of the segment.

$$\therefore \triangle = hk,$$
and $$J = \tfrac{2}{3} hk.$$

55. To find the area of an ellipse.

Rule: *Multiply the product of the semi-axes by π.*

Formula: $E = ab\pi$.

Proof: Let $ADA'D'$ be a circle of which AC, CD are radii at right angles to one another.

In CD let any point B be taken; then, if this point move so as to cut off from all ordinates of the circle the same part that BC is of DC, the curve traced is called an ellipse.

In one quadrant of the circle take a series of equidistant ordinates, as $Q_1 P_1 M_1$, $Q_2 P_2 M_2$, $Q_3 P_3 M_3$, etc. Draw $P_1 R_1$, $Q_1 R'_1$, etc., parallel to the axis AA'. Then, by 32,

area $R_1 M_1$: area $R'_1 M_1$: : $P_1 M_1$: $Q_1 M_1$: : $BC : AC$;

and each corresponding pair being in this constant ratio, $\therefore$ the sum of the rectangles RM is to the sum of $R'M$ as $BC : AC$. But the sum of RM differs from one-quarter of the ellipse by less than the area BM_1, which can be made less than any assignable quantity by taking CM_1, the common distance between the ordinates, sufficiently small.

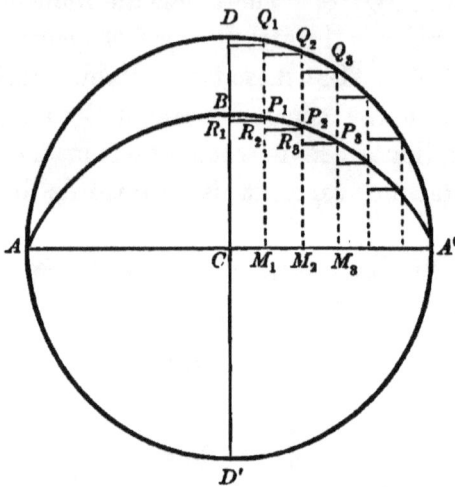

Hence, $A'BC$ is the limit of the sum of the rectangles RM; and, in the same way, the quadrant of the circle is the limit of the sum of $R'M$. Therefore, by 13,

$$\frac{E}{\odot} = \frac{BC}{AC} = \frac{b}{a}.$$
$$\therefore E = \frac{\odot b}{a} = \frac{a^2 \pi b}{a} = ab\pi.$$

Cor. The area of any segment of an ellipse, cut off by a line parallel to the minor axis, will be to the corresponding segment of the circle upon the major axis in the ratio of b to a.

EXAM. 52. Find the area of an ellipse whose major axis is 61·6 meters, and minor axis 44·4 meters.

$$E = 30·8 \times 22·2 \times 3·14159$$
$$= 2148·09+ \text{ square meters. } Ans.$$

CHAPTER IV.

The Measurement of the Areas of Broken and Curved Surfaces.

XVIII. A *polyhedron* is a solid bounded by polygons.

A polyhedron bounded by four polygons is called a *tetrahedron;* by six, a *hexahedron;* by eight, an *octahedron;* by twelve, a *dodecahedron;* by twenty, an *icosahedron*.

The *faces* of a polyhedron are the bounding polygons. If the faces are all congruent and regular, the polyhedron is *regular*.

The *edges* of a polyhedron are the lines in which its faces meet.

The *summits* of a polyhedron are the points in which its edges meet.

A *section* of a polyhedron is a polygon formed by the intersection of a plane with three or more faces.

A *convex* figure is such that a straight line cannot meet its boundary in more than two points.

⊁ 56. *The number of faces and summits in any polyhedron taken together exceeds by two the number of its edges.*

Formula: $\mathfrak{F} + \mathfrak{S} = \mathfrak{E} + 2$.

Proof: Let ϵ be any edge joining the summits $\alpha\beta$ and the faces AB, and let ϵ vanish by the approach of β to α. If A and B are neither of them triangles, they both remain, though reduced in rank and no longer collateral, and the figure has lost one edge ϵ and one summit β.

If B is a triangle and A no triangle, B vanishes with ϵ into an edge through a, but A remains. The figure has

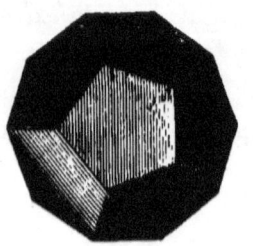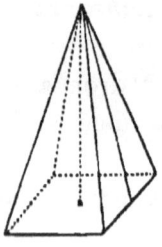

lost two edges of B, one face B, and one summit β. If B and A are both triangles, B and A both vanish with ϵ, five edges forming those triangles are reduced to two through a; and the figure has lost three edges, two faces, and the summit β.

In any one of these cases, whether one edge and one summit vanish, or two edges disappear with a face and a summit, or three edges with a summit and two faces, the truth or falsehood of the equation

$$\mathfrak{F} + \mathfrak{S} = \mathfrak{E} + 2$$

remains unaltered.

By causing all the edges which do not meet any face to vanish, we reduce the figure to a pyramid upon that face. Now, the relation is true of the pyramid; therefore it is true of the undiminished polyhedron.

§ (I). PRISM AND CYLINDER.

XIX. A *prism* is a polyhedron two of whose faces are congruent, parallel polygons, and the other faces are parallelograms.

The *bases* of a prism are the congruent, parallel polygons.

A *parallelepiped* is a prism whose bases are parallelograms.

A *normal* is a straight line perpendicular to two or more non-parallel lines.

The *altitude* of a prism is the normal distance between the planes of its bases.

A *right* prism is one whose lateral edges are normal to its bases.

57. To find the lateral surface or mantel of a prism.

Rule: *Multiply a lateral edge by the perimeter of a right section.*

Formula: $P = lp$.

Proof: The lateral edges of a prism are all equal.
The sides of a right section, being perpendicular to the

lateral edges, are the altitudes of the parallelograms which form the lateral area of the prism.

Cor. The lateral area of a *right* prism is equal to its altitude multiplied by the perimeter of the base.

EXAM. 53. The base of an oblique prism is a regular pentagon, each side being 3 meters, the perimeter of a right section is 12 meters, and the length of the prism 14 meters. Find the area of the whole surface.

By 46, the area of the pentagonal base is

$$9 \times 1{\cdot}7204774 = 15{\cdot}4842966.$$

Doubling this for the base and top together, and adding the lateral area of the prism, which, by 56, is

$$12 \times 14 = 168,$$

the total surface

$$= 168 + 30{\cdot}9685932 = 198{\cdot}9686-\text{ square meters. } Ans.$$

XX. A *cylindric* surface is generated by a straight line so moving that every two of its positions are parallel.

The generatrix in any position is called an *element* of the surface.

A *cylinder* is a solid bounded by a cylindric surface and two parallel planes.

The *axis* of a cylinder is the straight line joining the centers of its bases.

A *truncated* cylinder is the portion between the base and a non-parallel section.

58. To find the curved surface or mantel of a right circular cylinder.

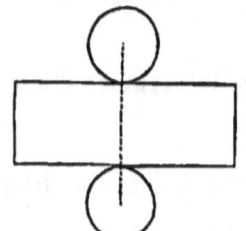

Rule: *Multiply its length by the circumference of its base.*

Formula: $C = cl = 2\pi rl$.

First Proof: Imagine the curved surface slit along an element and then spread out flat. It thus becomes a rectangle having for one side the circumference and for the adjacent side the length of the cylinder.

Second Proof: Inscribe in the right cylinder a right prism having a regular polygon as its base. Bisect the arcs subtended by the sides of this polygon, and thus inscribe a regular polygon of double the number of sides, and construct on it, as base, an inscribed prism.

Proceeding in this way continually to double the number of its sides, the base of the inscribed prism, by 14, approaches the base of the cylinder as its limit, and the prism itself approaches the cylinder as its limit. But, by 56,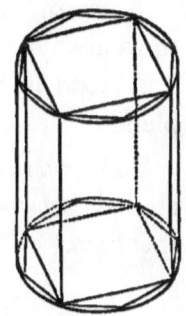

$$P = lp,$$

and always the variable P bears to the variable p the constant ratio l. Therefore, by 13, their limits are in the same ratio, and

$$C = cl.$$

Cor. 1. The curved surface of a truncated circular cylinder is the product of the circumference of the cylinder by the intercepted axis. For, by symmetry, substituting an oblique for a right section through the same point of the axis alters neither the curved surface nor the volume, since the solid between the two sections will be the same above and below the right section.

Cor. 2. The curved surface of any cylinder on any curve equals the length of the cylinder multiplied by the perimeter of a right section.

Exam. 54. Find the mantel of a right cylinder whose diameter is 18 meters and length 30 meters.

$C = 30 \times 18 \times 3{\cdot}14159 = 1696{\cdot}4586$ square meters. *Ans.*

§ (J). PYRAMID AND CONE.

XXI. A *regular pyramid* is contained by congruent isosceles triangles whose bases form a regular polygon.

A *conical* surface is generated by a straight line moving so as always to pass through a fixed point called the *vertex*.

A *cone* is a solid bounded by a conical surface and a plane.

The *frustum* of a pyramid or cone is the portion included between its base and a cutting plane parallel to the base.

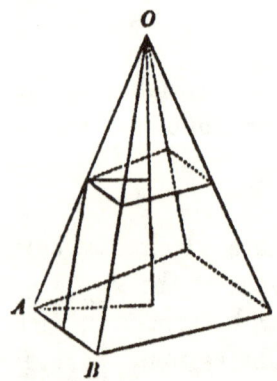

59. To find the area of the lateral surface or mantel of a regular pyramid.

Rule: *Multiply the perimeter of the base by half the slant height.*

Formula: $Y = \tfrac{1}{2} hp$.

Proof: The altitude of each of the equal isosceles triangles is the slant height of the pyramid, and the sum of their bases is the perimeter of its base.

EXAM. 55. Find the lateral area of a regular heptagonal pyramid whose slant height is 13·56224 meters, and basal edges each 1¼ meters.

One quarter of 13·56224 is 3·39056. Adding these and dividing their sum by 2 gives 8·4764 for the area of one triangular face. The lateral area is 7 times this, or

59·3348 square meters. *Ans.*

AREAS OF BROKEN AND CURVED SURFACES.

60. To find the area of the curved surface or mantel of a right circular cone.

Rule: *Multiply the circumference of its base by half the slant height.*

Formula: $K = \tfrac{1}{2} ch = \pi r h.$

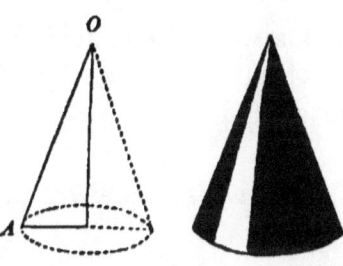

First Proof: The distance from the vertex of a cone of revolution to each point on the circumference of its base is the slant height of the cone. Therefore, if the surface of the cone be slit along a slant height and spread out flat, it becomes the sector of a circle, with the slant height as radius and the circumference of cone's base as arc.

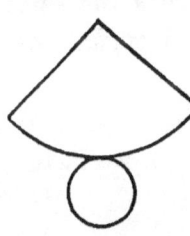

∴ by 48, its area is $\tfrac{1}{2} ch$.

Second Proof: About the base of the cone circumscribe a regular polygon, and join its vertices and points of contact to the vertex of the cone. Thus is circumscribed about the cone a regular pyramid whose slant height equals the slant height of the cone.

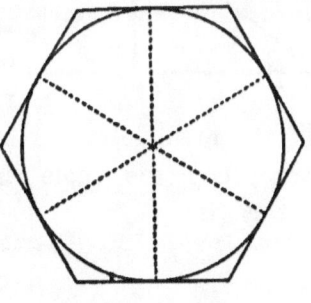

By drawing tangents, circumscribe a regular polygon of double the number of sides, and construct on it, as before, a circumscribed regular pyramid. Thus proceeding continually to double the number of sides, the base of the circumscribed pyramid, by 14, approaches the base of the cone as its limit, and the pyramid itself approaches the cone as its limit.

But, by 58,
$$Y = \tfrac{1}{2} hp,$$

and always the variable Y has to the variable p the constant ratio $\tfrac{1}{2} h$. ∴ by 13, their limits are in the same ratio, and
$$K = \tfrac{1}{2} ch.$$

Cor. 1. In the Proof of 47 we find
$$\odot = \tfrac{1}{2} cr.$$

∴ the slant height of a right circular cone has the same ratio to the radius of the base that the curved surface has to the base, or
$$K : B :: h : r.$$

Cor. 2. Calling λ the sector angle of the cone, we have
$$\lambda : 360 :: r : h.$$

EXAM. 56. Given the two sides of a right-angled triangle. Find the area of the surface described when the triangle revolves about its hypothenuse.

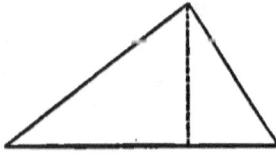

Calling a and b the given altitude and base, and x the length of the perpendicular from the right angle to the hypothenuse, by 59, the area described by a is $\pi x a$, and described by b is $\pi x b$. Thus the whole surface of revolution is $\pi(a+b)x$.
But
$$a : x :: \sqrt{a^2 + b^2} : b. \qquad \text{Eu. I. 47 \& VI. 8.}$$

$$\therefore x = \frac{ab}{\sqrt{a^2+b^2}}, \text{ and } \pi(a+b)x = \frac{\pi(a+b)ab}{(a^2+b^2)^{\frac{1}{2}}}. \quad Ans.$$

61. To find the lateral surface or mantel of the frustum of a regular pyramid.

AREAS OF BROKEN AND CURVED SURFACES.

Rule: *Multiply the slant height of the frustum by half the sum of the perimeters of its bases.*

Formula: $F = \tfrac{1}{2} h (p_1 + p_2).$

Proof: The base and top being similar regular polygons, the inclined faces are congruent trapezoids, the height of each being the slant height of the frustum. If n be the number of faces, by 40,

$$\text{area of each face} = \tfrac{1}{2} h \left(\frac{p_1}{n} + \frac{p_2}{n} \right).$$

$\therefore$ area of lateral surface $= F = \tfrac{1}{2} h (p_1 + p_2).$

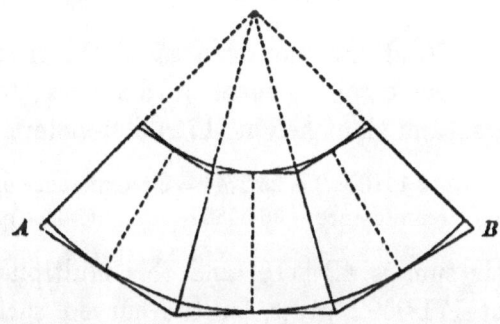

EXAM. 57. Find the lateral area of a regular pentagonal frustum whose slant height is 11·0382 meters, each side of its base being $2\tfrac{5}{6}$ meters, and of its top $1\tfrac{1}{2}$ meters.

The sum of a pair of parallel sides is $\tfrac{13}{3}.$

$$11\cdot0382 \times 13 = 143\cdot4966.$$
$$143\cdot4966 \div 6 = 23\cdot9161,$$

the area of one trapezoidal face. The lateral area is five times this, or

$$119\cdot5805 \text{ square meters. } Ans.$$

62. To find the curved surface or mantel of the frustum of a right circular cone.

Rule: *Multiply the slant height of the frustum by half the sum of the circumferences of its bases.*

Formula: $F = \tfrac{1}{2} h (c_1 + c_2) = \pi h (r_1 + r_2)$.

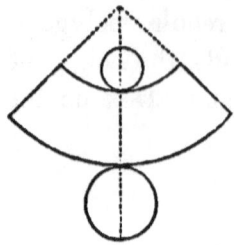

Proof: Completing the cone and slitting it along a slant height, the curved surface of the frustum develops into the difference of two similar sectors having a common angle, the arcs of the sectors being the circumferences of the bases of the frustum. By 53, the area of this annular sector $= F = \tfrac{1}{2} h (c_1 + c_2)$.

Exam. 58. Find the mantel area of the frustum of a right cone whose basal diameter is 18 meters; top diameter, 9 meters; and slant height, 171·0592 meters.

$3 \cdot 1416 \times 9 = 28 \cdot 2744 =$ circumference of top.
Twice top circumference $= 56 \cdot 5488 =$ circumference of base.

Half their sum is 42·4116, and this multiplied by the slant height 171·0592, gives, for the curved surface,

7254·89 square meters. *Ans.*

63. To find the curved surface of a frustum of a cone of revolution.

Rule: *Multiply the projection of the frustum's slant height on the axis by twice π times a perpendicular erected at the midpoint of this slant height and terminated by the axis.*

Formula: $F = 2 \pi a j$.

Proof: By 62, the curved surface of the frustum whose slant height is PR and axis MC is

$$F = \pi \times PR (PM + RN).$$

AREAS OF BROKEN AND CURVED SURFACES. 71

But, by 40, *Cor.*,
$$PM + RN = 2QO.$$
$$\therefore F = 2\pi \times PR \times QO.$$

But the triangle RPL is equiangular to CQO, since the three sides of one are perpendicular to the sides of the other.

$$\therefore PR \times QO = PL \times QC.$$

Ww. 279 & 259; (Eu. VI. 4 & 16; Cv. III. 25 & 5).

$$\therefore F = 2\pi(LP \times CQ) = 2\pi(MN \times CQ)$$
$$= 2\pi j a.$$

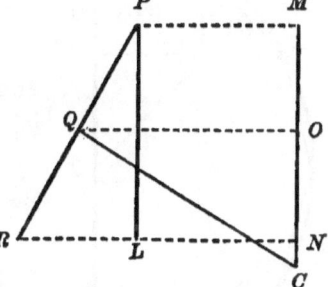

Cor. This remains true, if either PM or RN vanish, or if they become equal; that is, true for a cone or cylinder of revolution.

₹ (K). THE SPHERE.

XXII. A *sphere* is a closed surface all points of which are equally distant from a fixed point within called its *center*. A *globe* is the solid bounded by a sphere.

64. To find the area of a sphere.

Rule: *Multiply four times its squared radius by π.*

Formula: $H = 4r^2\pi$.

Proof: In a circle inscribe a regular polygon of an even number of sides. Then a diameter through one vertex passes through the opposite vertex, halving the polygon symmetrically. Let PR be one of its sides. Draw PM, RN perpendicular to the diameter BD. From the center C the perpendicular CQ bisects PR.

Ww. 183; (Eu. III. 3; Cv. II. 15).

Drop the perpendiculars PL, QO.

Now, if the whole figure revolve around BD as axis, the semicircumference will generate a sphere, while each side of the inscribed polygon, as PR, will generate the curved surface of the frustum of a cone. By 63, this

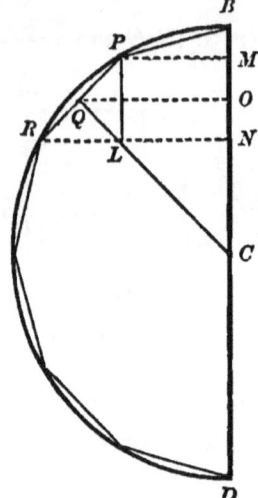

$$F = 2\pi MN \times CQ;$$

and the sum of all the frustums, that is, the surface of the solid generated by the revolving semi-polygon, equals $2\pi CQ$ into the sum of the projections.

$$\therefore \sum_{v=1}^{v=\frac{1}{2}n} F_v = 2\pi CQ \times BD = 2\pi a \times 2r = 4a r\pi.$$

As we double n the number of sides of the inscribed polygon, by 14, its semiperimeter approaches the semicircumference as limit, and its surface of revolution approaches the sphere as limit, while CQ or a, its apothem, approaches r the radius of the sphere as limit. But the variable sum bears to the variable a the constant ratio $4r\pi$. Therefore, by 13, their limits have the same ratio, and

$$H = 4r^2\pi.$$

Cor. 1. A sphere equals four times a circle with same radius.

Cor. 2. A sphere equals the curved surface of its circumscribing cylinder.

Exam. 59. Considering the earth as a sphere whose radius is $6·3709 \times 10^8$ centimeters, find its area.

$$H = 4(6·3709)^2 \times 10^{16} \times \pi.$$
$$H = 4 \times 40·58836681 \times 3·14159265 \times 10^{16}.$$

$H = 5{,}100{,}484{,}593{,}831{,}997{,}860$ square centimeters. *Ans.*
Or, about 510 million square kilometers.

AREAS OF BROKEN AND CURVED SURFACES. 73

XXIII. A *spherical segment* is the portion of a globe cut off by a plane, or included between two parallel planes.

A *zone* is the curved surface of a spherical segment.

The Proof of 64 gives also the following rule for the area of a zone:

65. To find the area of a zone.

Rule: *Multiply the altitude of the segment by twice π times the radius of the sphere.*

Formula: $Z = 2\pi r a$.

Cor. 1. Any zone is to the sphere as the altitude of its segment is to the diameter of the sphere.

Cor. 2. Let the arc BP generate a calot or zone of a single base. By 65, its area

$$Z = 2\pi \times BC \times BM = \pi BD \times BM = \pi \times BP^2.$$

Ww. 289; (Eu. VI. 8, Cor.; Cv. III. 44).

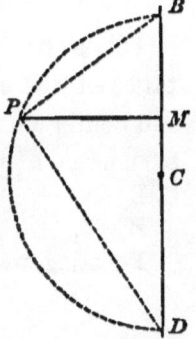

Hence, a calot or zone of one base is equivalent to a circle whose radius is the chord of the generating arc.

EXAM. 60. Find the area of a zone of one base, the diameter of this base being 60 meters, and the height of the segment 18 meters.

Using Cor. 2, the square of the chord of the generating arc is

$$(30)^2 + (18)^2 = 1224,$$

which, multiplied by π, gives, for the area of the calot,

3845·31 square meters. *Ans.*

X 66. **Theorem of Pappus.**

If a plane curve lies wholly on one side of a line in its own plane, and revolving about that line as axis generates thereby a surface of revolution, the area of the surface is equal to the product of the length of the revolving line into the path described by its center of mass.

Scholium. The demonstration given under the next rule, though fixing the attention on a single representative case, applies equally to all cases where the generatrix is a closed figure, has an axis of symmetry parallel to the axis of revolution, and so turns as to be always in a plane with the axis of revolution, while its points describe circles perpendicular to both axes.

Exam. 61. Use the Theorem of Pappus to find the distance of the center of mass of a semicircumference from the center of the circle, by reference to our formula for the surface of a sphere.

By 64, $H = 4r^2\pi$.
By 66, $H = r\pi \times 2x\pi$.
Equating, we get $2r = x\pi$.

$$\therefore x = \frac{2r}{\pi}. \text{ Ans.}$$

X 67. To find the area of the surface of a solid ring.

Rule: *Multiply the generating circumference by the path of its center.*

Formula: $O = 4\pi^2 r_1 r_2$.

Proof: Conceive any plane to revolve about any straight line in it. Any circle within the plane, but without the axis, will generate a solid ring.

AREAS OF BROKEN AND CURVED SURFACES. 75

Draw the diameter BCD parallel to the axis AO. Divide the semicircumference BPD into n equal arcs, and call their equal chords each k. From the points of division drop perpendiculars to the axis, thus dividing the

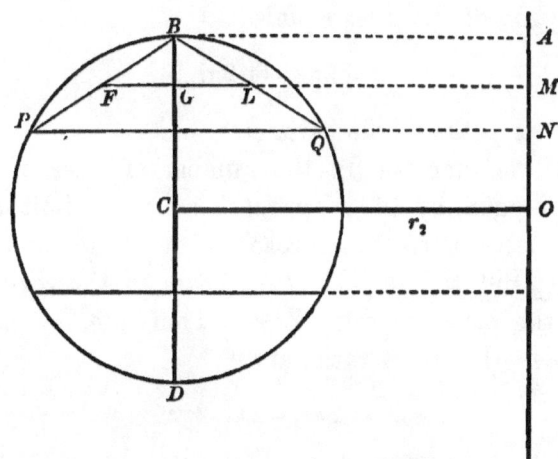

other semicircumference BQD into n corresponding parts. Let BP, BQ, be a pair of arcs. If we draw their chords, we have a pair of right-angled trapezoids, $ABPN$ and $ABQN$, which, during the revolution, describe frustums whose curved surfaces, by 62, are

$$F_1 = \pi k (BA + PN),$$
and $$F_2 = \pi k (BA + QN).$$

If $MLGF$ is the medial line, then, by Proof to 40,

$$F_1 = 2\pi k \times FM,$$
and $$F_2 = 2\pi k \times LM.$$

$\therefore F_1 + F_2 = 2\pi k (FM + LM) = 2\pi k (FG + GM + GM - GL).$

But the diameter BCD is an axis of symmetry, and

$$GM = CO = r_2,$$

the radius of the path of C.

$$\therefore F_1 + F_2 = 2\pi k 2r_2.$$

This is the expression for each pair; and, as we have n pairs, therefore, the whole surface generated by a symmetrical polygon of $2n$ sides equals

$$2nk\, 2\pi r_2 = p\, 2\pi r_2,$$

since $2nk$ is p the whole perimeter.

But, as we increase $2n$ the number of sides of the inscribed polygon, by 14, p approaches c as its limit, and the sum of frustral surfaces approaches the surface of the ring as limit. But the variable sum bears to the variable perimeter the constant ratio $2\pi r_2$. Therefore, by 13, their limits have the same ratio, and

$$O = c\, 2\pi r_2 = 2\pi r_1\, 2\pi r_2,$$

where r_1 is the radius of the generating circle, and r_2 the radius of the path of its center.

Exam. 62. Find the surface of a solid ring, of which the thickness is 3 meters, and the inner diameter 8 meters.
Here r_1 is $1\tfrac{1}{2}$ meters, and r_2 is $5\tfrac{1}{2}$ meters.

$$\therefore 4\pi^2 r_1 r_2 = 3.1416 \times 3 \times 3.1416 \times 11$$
$$= 9.4248 \times 34.5576$$
$$= 325.698 \text{ square meters. } Ans.$$

Exam. 63. Find the area of the surface of a square ring described by a square meter revolving round an axis parallel to one of its sides, and 3 meters distant.
Here the length of the generating perimeter is 4 meters. The path of its center is 7π, since r_2 is $3\tfrac{1}{2}$ meters.

$$\therefore O = 28\pi = 87.9648 \text{ square meters. } Ans.$$

AREAS OF BROKEN AND CURVED SURFACES. 77

EXAM. 64. A circle of 1·35 meters radius, with an inscribed hexagon, revolves about an axis 6·25 meters from its center and parallel to a side of the hexagon. Find the difference in area of the generated surfaces.

Here
$$r_1 = 1·35 \quad \text{and} \quad r_2 = 6·25.$$

Therefore, area of circular ring is

$$4\pi^2 r_1 r_2 = \pi^2 \times 2·7 \times 12·5 = 9·8696 \times 33·75 = 333·099.$$

For the hexagonal ring

Ww. 391; (Eu. IV. 15, Cor.; Cv. V. 14)

the length of the generating perimeter is

$$6 \times 1·35 = 8·1.$$

The path of its center is

$$\pi \times 12·5 = 39·27-.$$

Therefore, its area is

$$39·27 \times 8·1 = 318·087.$$

Thus the difference in area is

15·012 square meters. *Ans.*

X §(L). SPHERICS AND SOLID ANGLES.

XXIV. A *great circle* is a section of a globe made by a plane passing through the center.

A *lune* is that portion of a sphere comprised between two great semicircles.

The *angle of two curves* passing through the same point is the angle formed by the two tangents to the curves at that point.

A *spherical angle* is the angle included between two arcs of great circles.

A *plane angle* is the amount of divergence between two straight lines which meet in a point.

A *solid angle* is the amount of spread between two or more planes which meet at a point.

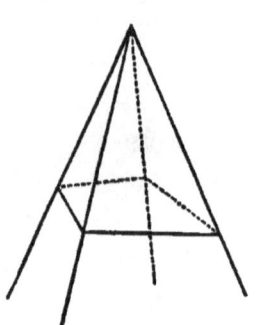

Two polyhedral angles, having all their parts congruent, but arranged in reverse order, are *symmetrical*.

A *steregon*, the natural unit of solid angle, is the whole amount of solid angle round about a point in space.

As a perigon corresponds to a circle and its circumference, so a steregon corresponds to a globe and its sphere.

The steregon is divided into 360 equal parts, called spherical degrees of angle, and these divide the whole sphere into 360 equal parts, each called a degree of spherical surface.

A *steradian* is the angle subtended at the center by that part of every sphere equal to the square of its radius.

68. To find the area of a lune.

Rule: *Multiply its angle in radians by twice its squared radius.*

Formula: $L = 2r^2 u$.

Proof: Let $PAQBP$, $PBQCP$ be two lunes having equal angles at P; then one of these lunes may be placed on the other so as to coincide exactly with it: thus *lunes having equal angles are congruent*. Then, by the process of Eu. VI. 33; (Ww. 766; Cv. VIII. 95), it follows that *a lune is to the sphere as its angle is to a perigon;*

$$\frac{L}{4r^2\pi} = \frac{u}{2\pi}; \qquad \therefore L = 2r^2 u.$$

Cor. 1. A lune contains as many degrees of spherical surface as its angle contains degrees.

Cor. 2. A lune measures twice as many steradians as its angle contains radians.

EXAM. 65. Find the area comprised between two meridians one degree apart on the earth's surface.

Assuming as the earth's surface 196,625,000 square miles, dividing by 360, gives for the lune,

$$546{,}180{\cdot}5+ \text{ square miles. } Ans.$$

XXV. Suppose the angular point of a solid angle is made the center of a sphere; then the planes which form the solid angle will cut the sphere in arcs of great circles.

Thus a figure will be formed on the sphere, which is called a *spherical triangle* if it is bounded by *three* arcs of great circles, each less than a semicircumference.

If the solid angle be formed by the meeting of *more than three* planes, the corresponding figure on the sphere is bounded by more than three arcs of great circles, and is called a *spherical polygon*.

The solid angle made by only two planes corresponds to the lune intercepted on any sphere whose center is in the common section of the two planes.

The *dihedral* angle of two planes is the amount of rotation which one plane must make about their intersection in order to coincide with the other.

The angles of a spherical polygon equal the dihedral angles of its solid angle.

The sides of the polygon measure the *face angles* of this polyhedral angle.

From any property of polyhedral angles we may infer an analogous property of spherical polygons. Reciprocally, from any property of spherical polygons we may infer a corresponding property of polyhedral angles.

XXVI. A *spherical pyramid* is a portion of a globe bounded by a spherical polygon and the planes of the sides of the polygon.

The center of the sphere is the *vertex* of the pyramid; the spherical polygon is its *base*.

69. Just as plane angles at the center of a circle are proportional to their intercepted arcs, and also sectors; so *solid angles at the center of a sphere are proportional to their intercepted spherical polygons, and also spherical pyramids.*

XXVII. The *spherical excess* of a spherical triangle is the excess of the sum of its angles over a flat angle. The spherical excess of a spherical polygon is the excess of the sum of its angles above as many flat angles as it has sides less two.

70. To find the area of a spherical triangle.

Rule: *Multiply its spherical excess in radians by its squared radius.*

Formula: $\widehat{\triangle} = er^2$.

Proof: Let ABC be a spherical triangle. Produce the arcs which form its sides until they meet again, two and two. The $\widehat{\triangle} ABC$ now forms a part of three lunes, namely, $ABDCA$, $BCEAB$, and $CAFBC$.

Since the $\widehat{\triangle}$'s CDE and FAB subtend vertical solid angles at O, they are equivalent, by 69. Therefore, the lune $CAFBC$ equals the sum of the two triangles ABC and CDE. Thus the lunes whose angles are A, B, and C, are together equal to a hemisphere plus twice $\widehat{\triangle}\ ABC$. Subtracting the hemisphere, which equals a lune whose angle is a flat angle, we have

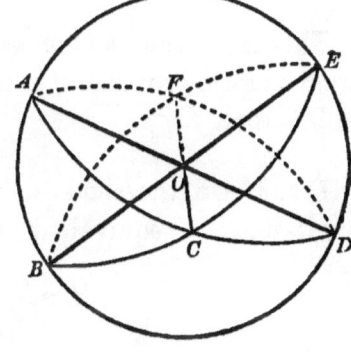

$$2\widehat{\triangle}\ ABC =$$
lune whose ∡ is $(A + B + C - f)$.

∴ $\widehat{\triangle}$ = lune whose ∡ is $\tfrac{1}{2}e$.

∴ by 68, $\widehat{\triangle} = er^2$.

Cor. 1. A $\widehat{\triangle}$ contains half as many degrees of spherical surface as its e contains degrees.

Cor. 2. A $\widehat{\triangle}$ measures as many steradians as its e contains radians.

Cor. 3. Every ∡ of a $\widehat{\triangle}$ is $> \tfrac{1}{2} e$.

EXAM. 66. Find the area of a tri-rectangular $\widehat{\triangle}$.
Here
$$e = \text{rt. ∡} = \tfrac{1}{2}\pi.$$
$$\therefore \widehat{\triangle} = \tfrac{1}{2}\pi r^2;$$

or, a tri-rectangular triangle is one-eighth of its sphere.

By Cor. 1, a tri-rectangular $\widehat{\triangle}$ contains forty-five degrees of spherical surface.

71. To find the area of a spherical polygon.

Rule: *Multiply its spherical excess in radians by its squared radius.*

Formula: $\widehat{N} = [\hat{s} - (n-2)\pi] r^2$.

Proof: From any angular point divide the polygon into $(n-2)$ $\widehat{\triangle}$'s. ∴ by 70,

$$\widehat{N} = [\mathfrak{k} - (n-2)\pi]r^2.$$

This expression is true even when the polygon has reëntrant angles, provided it can be divided into $\widehat{\triangle}$'s with each $\measuredangle$ less than f.

Cor. 1. On the same or equal spheres, n-gons of equal angle-sum are equivalent; or,

$$\widehat{N}_1 = \widehat{N}_2, \quad \text{if} \quad \mathfrak{k}_1 = \mathfrak{k}_2.$$

Cor. 2. To construct a dihedral solid $\measuredangle$ equal to any polyhedral $\measuredangle$; that is, to transform into a lune any spherical polygon; add its angles, subtract $(n-2)f$, and halve the remainder.

EXAM. 67. Find the ratio of the vertical solid angles of two right cones of altitude a_1 and a_2, but having the same slant height h.

These solid angles are as the corresponding calots on the sphere of radius h.

Therefore, from 65, the required ratio is

$$\frac{2\pi h(h-a_1)}{2\pi h(h-a_2)} = \frac{h-a_1}{h-a_2},$$

the ratio of the calot-altitudes. For the equilateral and right-angled cones this becomes

$$\frac{2-\sqrt{3}}{2-\sqrt{2}}.$$

Third Reference Table of Abbreviations.

$\left.\begin{array}{l}\alpha\\\beta\\\gamma\end{array}\right\}$ = angles.

δ = density.
ϵ = edge.
ζ = V. of paraboloid.
η = V. of ellipsoid.
θ = V. of prolate spheroid.
ι = $\sqrt{-1}$.
λ = $\angle$ of cone.
μ = mass.
ξ = approximation.

π = $\dfrac{c}{d}$.
ρ = radian.
σ = V. of oblate spheroid.
τ = distance.
$\hat{\upsilon}$ = V. of spherical ungula.
ϕ = function.
χ = V. of hyperboloid.
ψ = V. of mid F. of spindle.
ω = weight.
$\propto$ = varies as.
$\cong$ = congruent.

$\doteq$ = approaches.
μC = mass-center.

CHAPTER V.

THE MEASUREMENT OF VOLUMES.

⸙ (M). PRISM AND CYLINDER.

XXVIII. Two polyhedrons are *symmetrical* whose faces are respectively congruent, and whose polyhedral angles are respectively symmetrical; *c.g.*, a polyhedron is symmetrical to its image in a mirror.

A *quader* is a parallelepiped whose six faces are rectangles.

A *cube* is a quader whose six faces are squares.

XXIX. The *volume* of a solid is its ratio to an assumed unit.

The *unit* for measurement *of volume* is a cube whose edge is the unit of length.

Thus, if the linear unit be a meter, the unit of volume, contained by three square meters at right angles to each other, is called a *cubic meter* (^{cbm}).

XXX. Using *length* of a line to mean its numerical value, lengths, areas, and volumes, are all three quantities of the same kind, namely *ratios*. All ratios, whether expressible as numbers or not, combine according to the same simple laws as ordinary numbers and fractions.

Ww. Bk. III.; (Eu. Bk. V.; Cv. Bk. II.).

THE MEASUREMENT OF VOLUMES.   85

Therefore, we may multiply lengths, areas, and volumes together promiscuously, or divide one by the other in any order.

If we ever speak of multiplying a line, a surface, or a solid, we mean always the length of the line, the area of the surface, or the volume of the solid.

72. To find the volume of a quader.

Rule: *Multiply together the length, breadth, and height of the quader.*

Or, in other words,

Multiply together the lengths of three adjacent edges.

Formula: $U = abl$.

Proof: By 32, the number of square units in the base of a quader is the product of two adjacent edges, bl.

If on each of these square units we place a unit cube, for every unit of altitude we have a layer of bl cubic units; so that, if the altitude is a, the quader contains abl cubic units.

Cor. 1. The volume of any cube is the third power of the length of an edge; and this is why the third power of a number is called its cube.

Cor. 2. Every unit of volume is equivalent to a thousand of the next lower order.

Cor. 3. The arithmetical or algebraic extraction of cube root makes familiar the use of the equation

$$(a + b)^3 = a^3 + 3a^2b + 3ab^2 + b^3.$$

Its geometric meaning and proof follow from inspection of the figure of a cube on the edge $a+b$, cut by three planes into eight quaders.

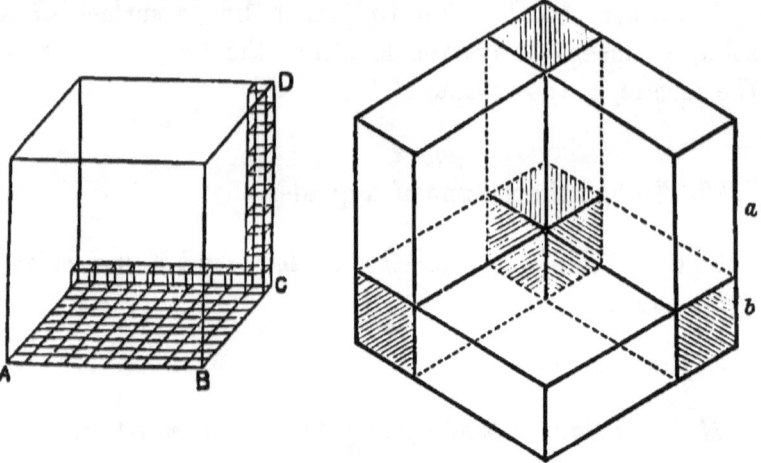

The cube a^3 of the longer rod a, taken out, had faces a^2 in common with three quaders of altitude b; had edges a in common with three quaders of base b^2, and one corner the corner point of the smaller cube b^3.

XXXI. Mass, Density, Weight.

The *unit of capacity* is a cubic decimeter, called the *liter* ([1]).

The *quantity of matter* in a body is termed its *mass*. The *unit of mass* is called a *gram* ([g]). Pure water at temperature of maximum density is 1·000013 gram per cubic centimeter (^{ccm}). So, in physics, the centimeter is chosen as the unit of length, because of the advantage of making the *unit of mass* practically identical with the *mass of unit-volume of water;* in other words, of making the value of the *density of water* practically equal to *unity; density* being defined as *mass per unit-volume*.

The *second* is the fundamental *unit of time* adopted with the centimeter and the gram.

Though the *weight of a body*, that is, the force of its attraction toward the earth, *varies* according to locality, yet weight being proportional to mass, the number expressing the mass of a body expresses also its weight in terms of the weight of the mass-unit at the same place. Thus, *in terms*

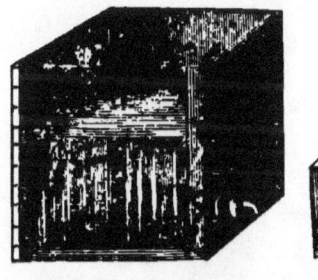
Liter = Cubic Decimeter.

Cubic Centimeter.

Gram Weight.
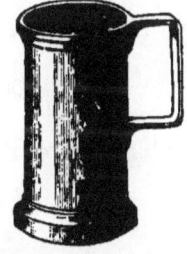
Liter (common form).

of the gram and centimeter, or of the kilogram (^{kg}) and liter, the *mass, weight, and volume of water* are expressed by the *same number.*

So the *density of any substance* is the *number of times* the *weight of the substance* contains the *weight* of an *equal bulk of water.* Therefore, the *density* of a substance is the *weight* of a *cubic centimeter* of that substance in *grams*, or the weight of a *liter* in *kilograms.* Hence,

73. To find the density of a body.

Rule: *Divide the weight in grams by the bulk in cubic centimeters.*

Formula: $\delta = \dfrac{\omega^g}{V^{ccm}} = \dfrac{\omega^{kg}}{V^l}.$

Exam. 68. If 65 cubic centimeters of gold weigh 1251·77 grams; find its density.
$$1251·77 \div 65 = 19·258. \ Ans.$$

Exam. 69. How many cubic centimeters (^{ccm}) in one hektoliter (^{hl})?

Since 1 liter = 1000 cubic centimeters,

$\therefore$ 1 hektoliter = 100,000 cubic centimeters. *Ans.*

Exam. 70. If the density of iron is 7·788, find the mass of a rectangular iron beam 7 meters long, 25 centimeters broad, and 55 millimeters high.

The volume of the beam in cubic centimeters is

$$700 \times 25 \times 5·5 = 96,250 \text{ cubic centimeters.}$$

Therefore, its mass is

$$96,250 \times 7·788 = 749,595 \text{ grams. } \textit{Ans.}$$

74. To find the volume of any parallelepiped.

Rule: *Multiply its altitude by the area of its base.*

Formula: V. P $= abl$.

Proof: Any parallelepiped is equivalent to a quader of equal base and altitude. For, supposing AB an oblique parallelepiped on an oblique base, prolong the four edges parallel to AB, and cut them normally by two parallel planes whose distance apart, CD, is equal to AB. This gives us the parallelepiped CDE, which is still oblique, but on a rectangular base. Prolong the four edges parallel to DE, and cut them normally by two planes whose distance apart FG is equal to DE. This gives us the quader FG.

Now, the solids AC and BDE are congruent, having all their angles and edges respectively equal. Subtracting each in turn from the whole solid ADE leaves CDE equivalent to AB.

Again, the solids CDF and EG are congruent. Taking from each the common part EF leaves CDE equivalent

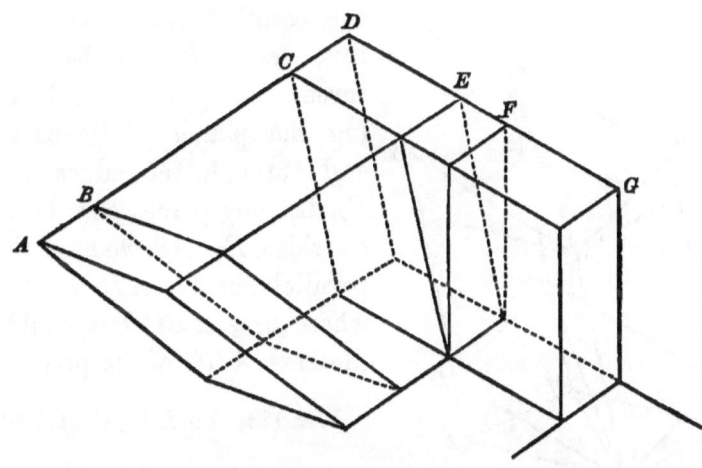

to FG. Therefore, the parallelepiped AB is equivalent to the quader FG of equal base and altitude.

EXAM. 71. The square of the altitude of a parallelepiped is to the area of its base as 121 to 63, and it contains 1,901,592 cubic centimeters. Find its altitude.

Here
$$aB = 1,901,592 \text{ and } 63\,a^2 = 121\,B.$$
$$\therefore 63\,a^3 = 121 \times 1,901,592 = 230,092,632.$$
$$\therefore a^3 = 230,092,632 \div 63 = 3,652,264.$$

$$\therefore a = 154 \text{ cubic centimeters. } Ans.$$

75. To find the volume of any prism.

Rule: *Multiply the altitude of the prism by the area of its base.*

Formula: V. $P = aB$.

Proof: For a three-sided prism this rule follows from 74, since any three-sided prism is half a parallelepiped of the same altitude, the base of the prism being half the base of this parallelepiped. To show this, let $ABCαβγ$ be any three-sided prism. Extending the planes of its bases, and through the edges $Aα$, $Cγ$, drawing planes parallel to the sides, $Bγ$, $Aβ$, we have the parallelepiped $ABCDαβγδ$, whose base $ABCD$ is double the base ABC of the prism.

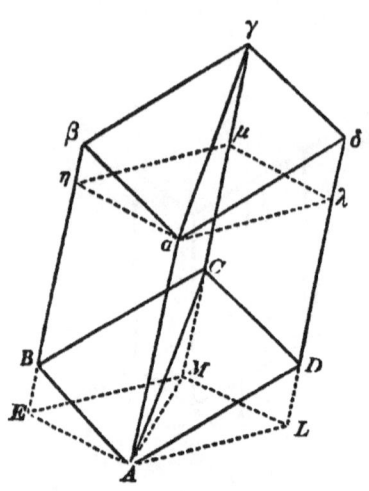

Ww. 123; (Eu. I. 34; Cv. I. 105).

Also, this parallelepiped itself is twice the prism. For, its two halves, the prisms, are congruent if its sides are all rectangles. If not, the prisms are symmetrical and equivalent. For, draw planes perpendicular to $Aα$ at the points A and $α$. Then the prism $ABCαβγ$ is equivalent to the right prism $AEMαημ$, because the pyramid $AEBCM$ is congruent to the pyramid $αηβγμ$. In the same way, $ADCαδγ$ equals $ALMαλμ$.

But $AEMαημ$ and $ALMαλμ$ are congruent. Therefore, $ABCαβγ$ and $ADCαδγ$ are equivalent, and the parallelepiped $ABCDαβγδ$ is double the prism $ABCαβγ$.

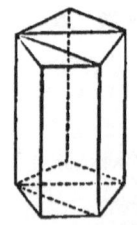

Thus, the rule is proved true for triangular prisms, and consequently for all prisms; since, by passing planes through any one lateral edge, and all the other lateral edges, excepting the two adjacent, we can divide any prism into a number of triangular prisms of the same altitude, whose triangular bases together make the given polygonal base.

THE MEASUREMENT OF VOLUMES. 91

Cor. 1. The volume of any prism equals the product of a lateral edge by the cross-section normal to it.

Cor. 2. Every parallelepiped is halved by each diagonal plane.

Cor. 3. Every plane passing through *two* opposite corners, halves the parallelepiped.

Cor. 4. The volume of a truncated parallelepiped equals half the sum of two opposite lateral edges multiplied by the cross-section normal to them.

EXAM. 72. The altitude of a prism is 5 meters, and its base a regular triangle. If, with density 4, it weighs 1836 kilogrammes, find a side of its base.

Its volume is 1836 ÷ 4 = 449 cubic decimeters.
The area of its base = 459 ÷ 50 = 9·18 square decimeters.

By 36, *Cor.*, the square of a side of this regular triangle is

9·18 ÷ 0·433 = 21·2 square decimeters.

Therefore, a side equals   4·604+ decimeters. *Ans.*

76. To find the volume of any cylinder.

Rule: *Multiply the altitude of the cylinder by the area of its base.*

Formula when Base is a Circle: $V. C = ar^2\pi$.

Proof: In 58, Second Proof, we saw the cylinder to be the limit of an inscribed prism when the number of sides of the prism is increased indefinitely, and the breadth of each side indefinitely diminished, the base of the cylinder being consequently the limit of the base of the prism. But, by 75, always V. P is to B in the constant ratio a; hence, by 13, their limits will be to one another in the same ratio; and V. C $= aB$.

Scholium. This applies to all solids whose cross-section does not vary, whatever be the shape of the cross-section.

Cor. 1. Between any two parallel planes, the volume of any cylinder equals the product of its axis by the cross-section normal to it.

Cor. 2. By 58, Cor. 1, the volume of any truncated circular cylinder equals the product of its axis by the circle normal to it.

EXAM. 73. A gram of mercury, density 13·6, fills a cylinder 12 centimeters long; find the diameter of the cylinder.

Volume of cylinder $= \frac{1}{13\cdot 6} =$ ·073529+ cubic centimeters.
Area of base of cylinder $= r^2\pi =$ ·073529 ÷ 12
$=$ ·0061274$\frac{2}{3}$ square centimeters.
Therefore $r^2 =$ ·0061274$\frac{2}{3}$ ÷ 3·1416 $=$ ·001950.
Thus
and $\qquad r =$ ·044 centimeters $=$ ·44 millimeters,
$\qquad d = 2r =$ ·88 millimeters. *Ans.*

77. To find the volume of a cylindric shell.

Rule: *Multiply the sum of the inner and outer radii by their difference, and this product by π times the altitude of the shell.*

THE MEASUREMENT OF VOLUMES.

Formula: $V. C_1 - V. C_2 = a\pi(r_1 + r_2)(r_1 - r_2)$.

Proof: Since a cylindric shell is the difference between two circular cylinders of the same altitude, its volume equals

$$ar_1^2\pi - ar_2^2\pi = a\pi(r_1^2 - r_2^2).$$

EXAM. 74. The thickness of the lead in a pipe weighing 94·09 kilograms is 6 millimeters, the diameter of the opening is 4·8 centimeters; taking $\pi = \tfrac{22}{7}$, and density 11, find the length of the pipe.

Here $r_2 = 2·4$ centimeters and $r_1 = 3$ centimeters.

$$94{,}090 = 11\, l\pi (r_1 + r_2)(r_1 - r_2)$$
$$= 11\, l\tfrac{22}{7} 5·4 \times ·6$$
$$= \tfrac{242}{7} l\, 3·24$$
$$= \tfrac{1}{1} l\, 784·08;$$
$$\therefore\ 658{,}630 = 784·08\, l.$$

$$\therefore\ l = 658{,}630 \div 784·08 = 840 \text{ centimeters}$$
$$= 8·4 \text{ meters.} \qquad Ans.$$

§ (N). PYRAMID AND CONE.

XXXII. The altitude of a pyramid is the normal distance from its vertex to the plane of its base.

78. Parallel plane sections of a pyramid are similar figures, and are to each other as the squares of their distances from the vertex.

Proof: The figures are similar, since their angles are respectively equal, Ww. 462; (Eu. XI. 10; Cv. VI. 32). and their sides proportional.

Ww. 279; (Eu. VI. 4; Cv. III. 25).

By 44, they are to each other as the squares of homologous sides, and hence as the squares of the normals from the vertex.

Ww. 469; (Eu. XI. 17; Cv. VI. 37).

Scholium. This is the reason why the strength of gravity, light, heat, magnetism, electricity, and sound, decreases as the square of the distance from the source.

Image part of the beams from a luminous point as a pyramid of light. If a cutting plane is moved away parallel to itself, the number of units of area illuminated increases as the square of the distance. But the number of rays remains unchanged. Therefore, the number of beams striking a unit of area must decrease as the square of the distance.

79. Tetrahedra (triangular pyramids) having equivalent bases and equal altitudes are equivalent.

Proof: Divide the equal altitudes a into n equal parts, and through each point of division pass a plane parallel to the base. By 78, all the sections in the first tetrahedron are triangles equivalent to the corresponding sections in the second.

Beginning with the base of the first tetrahedron, construct on each section as lower base a prism $\frac{a}{n}$ high with lateral edges parallel to one of the edges of the tetrahedron.

In the second, similarly construct prisms on each section as upper base.

Since the first prism-sum is greater than the first tetrahedron, and the second prism-sum less than the second tetrahedron, therefore the difference of the tetrahedra is less than the difference of the prism-sums.

But, by 75, each prism in the second tetrahedron is equivalent to the prism next above it on the first tetrahedron.

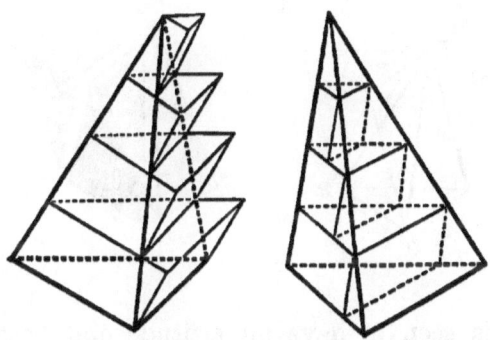

So the difference of the prism-sums is simply the lowest prism of the first series, whose volume, by 75, is $\frac{aB}{n}$.

As n increases this decreases, and can be made less than any assignable quantity by taking n sufficiently great. Hence the tetrahedra can have no assignable difference; and, being constants, they cannot have a variable difference.

Therefore the tetrahedra are equivalent.

Scholium. This demonstration indicates a method of proving that any two solids having equivalent bases and equal altitudes are equivalent, if every two plane sections at the same distance from the base are equivalent.

80. To find the volume of any pyramid.

Rule: *Multiply one-third of its altitude by the area of its base.*

Formula: $V.Y = \frac{1}{3}aB$.

Proof: Any triangular prism, as ABC–FDE, can be divided into three tetrahedra, two (B–DEF and D–ABC) having the same altitude as the prism, and its top and bottom respectively as bases, while the third

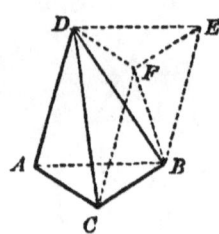 

($BCDF$) is seen to have an altitude and base equal to each of the others in turn by resting the prism first on its side CE and next on its side AF. Hence, by 79, these three tetrahedra are equivalent, and therefore, by 75, the volume of each is $\tfrac{1}{3}aB$.

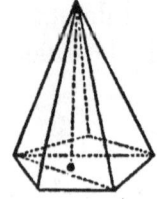

The rule thus proved for triangular pyramids is true for all pyramids, since, by passing planes through any one lateral edge, and all the other lateral edges excepting the two adjacent to this one, we can exhibit any pyramid as a sum of tetrahedra having the same altitude whose bases together make the given polygonal base.

EXAM. 75. If the altitude of the highest Egyptian pyramid is 138 meters, and a side of its square base 228 meters, find its volume.

Here
$$V.\,Y = \tfrac{1}{3} 138 (228)^2$$
$$= 46 \times 51{,}984$$
$$= 2{,}391{,}264 \text{ cubic meters. } Ans.$$

81. To find the volume of any cone.

THE MEASUREMENT OF VOLUMES. 97

Rule: *Multiply one-third its altitude by the area of its base.*

Formula when Base is a Circle: $V. K = \frac{1}{3} a r^2 \pi$.

Proof: In 60, Second Proof, we saw that the base of a cone was the limit of the base of the circumscribed or inscribed pyramid, and therefore the cone itself the limit of the pyramid. But, by 80, always the variable pyramid is to its variable base in the constant ratio $\frac{1}{3} a$.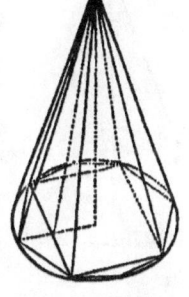
Therefore, by 13, their limits are to one another in the same ratio and

$$V. K = \tfrac{1}{3} a B.$$

Scholium. This applies to all solids determined by an elastic line stretching from a fixed point to a point describing any closed plane figure.

Cor. The volume of the solid generated by the revolution of any triangle about one of its sides as axis is one-third the product of the triangle's area into the circumference described by its vertex.

$$V = \tfrac{2}{3} \pi r \triangle.$$

EXAM. 76. Find the volume of a conical solid whose altitude is 15 meters and base a parabolic segment 3 meters high from a chord 11 meters long.

By 54, here
$$\begin{aligned} V. K &= \tfrac{1}{3} 15 \times \tfrac{2}{3} 3 \times 11 \\ &= 5 \times 2 \times 11 \\ &= 110 \text{ cubic meters. } Ans. \end{aligned}$$

98   MENSURATION.

EXAM. 77. Required the volume of an elliptic cone, the major axis of its base being 15·2 meters; the minor axis, 10 meters; and the altitude, 22 meters.

By 55, here

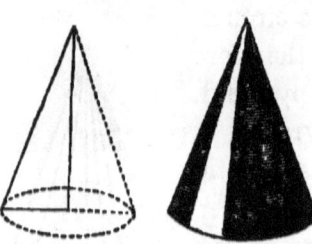

$$V. K = \tfrac{22}{3} 7\cdot 6 \pi 5$$
$$= 38\pi 7\tfrac{1}{3}$$
$$= \pi 278\tfrac{2}{3}$$
$$= 875\cdot 45+ \text{ cubic meters. } Ans.$$

EXAM. 78. The section of a right circular cone by a plane through its vertex, perpendicular to the base is an equilateral triangle, each side of which is 12 meters; find the volume of the cone.

Here

$$a = \sqrt{12^2 - 6^2} = \sqrt{108}.$$

$$\therefore \tfrac{1}{3} a r^2 \pi = \tfrac{1}{3}\sqrt{108}\,\pi\, 36$$
$$= 391\cdot 78 \text{ cubic meters. } Ans.$$

§ (O). **PRISMATOID.**

XXXIII. If, in each of two parallel planes is constructed a polygon, in the one an *m*-gon, *e.g.*, $ABCD$; in the other, an *n*-gon, *e.g.*, $A'B'C'$; then, through each side of one and each vertex of the other polygon a plane may be passed.

Thus, starting from the side AB, we get the *n* planes ABA', ABB', ABC'; again, with the side BC, the *n* planes BCA', BCB', BCC', etc. Using thus all *m* sides of the polygon $ABCD$, we get mn planes. Also, combining each of the *n* sides $A'B'$, $B'C'$, $C'D'$ with each of the *m* points A, B, C, D, gives nm planes $A'B'A$, $A'B'B$,

$A'B'C$, $A'B'D$; $B'C'A$, $B'C'B$, etc.; so that, altogether, $2mn$ connecting planes are determined by the two polygons. Among these are $m+n$ outer planes, which to-

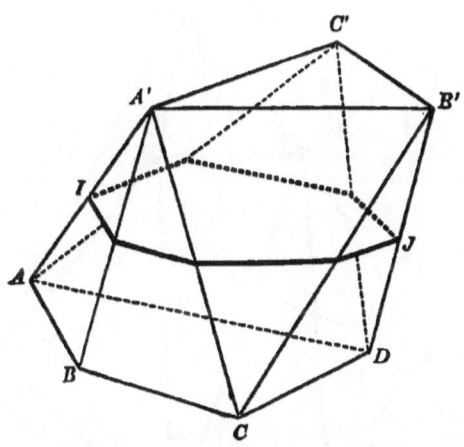

gether enclose the rest. These outer planes form the sides, and the given polygons the bases of a solid called a *prismatoid*. Our figure is a case of this body when

$$m=4 \text{ and } n=3.$$

The midcross-section IJ is given to show the seven sides.

XXXIV. A *prismatoid* is a polyhedron whose bases are any two polygons in parallel planes, and whose lateral faces are determined by so joining the vertices of these bases that each line in order forms a triangle with the preceding line and one side of either base.

REMARK. This definition is more general than XXXIII., and allows dihedral angles to be concave or convex, though neither base contain a reëntrant angle. Thus, BB' might have been joined instead of $A'C$.

From the prismatoids thus pertaining to the same two bases, XXXIII. chooses the greatest.

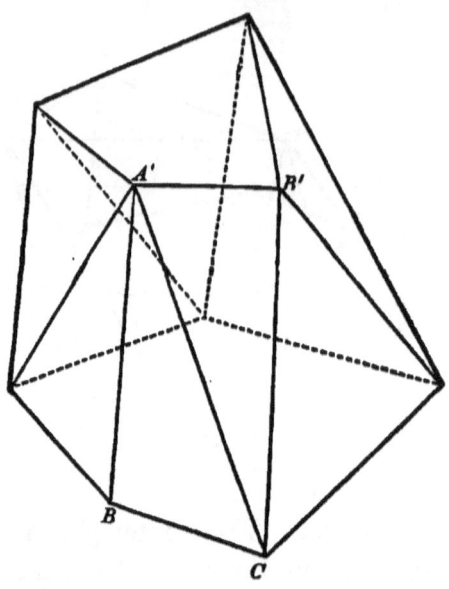

XXXV. The *altitude* of a prismatoid is the normal distance between the planes of its bases. Passing through the middle point of the altitude a plane parallel to the bases gives the midcross-section. Its vertices halve the lateral edges of the prismatoid. Hence, its perimeter is half the sum of the basal perimeters. But, if one base reduces to a straight line, this line must be considered a digon, *i.e.*, counted twice.

XXXVI. In stereometry the prism, pyramid, and prismatoid correspond respectively to the parallelogram, triangle, and trapezoid in planimetry.

XXXVII. Though, in general, the lateral faces of a prismatoid are triangles, yet if two basal edges which form,

THE MEASUREMENT OF VOLUMES. 101

with the same lateral edge, two sides of two adjoining faces are parallel, then these two triangular faces fall in the same plane, and together form a trapezoid.

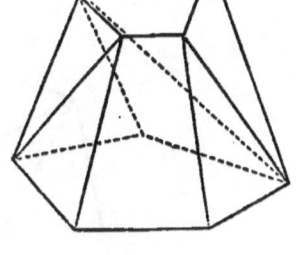

XXXVIII. A *prismoid* is a prismatoid whose bases have the same number of sides, and every corresponding pair parallel.

XXXIX. A *frustum of a pyramid* is a prismoid whose two bases are similar.

Cor. Every three-sided prismoid is the frustum of a pyramid.

XL. If both bases of a prismatoid become lines, it is a tetrahedron.

XLI. A *wedge* is a prismatoid whose lower base is a rectangle, and upper base a line parallel to a basal edge.

82. To find the volume of any prismatoid.

Rule: *Add the areas of the two bases and four times the midcross-section; multiply this sum by one-sixth the altitude.*

Prismoidal Formula: $D = \tfrac{1}{6} a (B_1 + 4 M + B_2)$.

Proof: In the midcross-section of the prismatoid take a point N, which join to the corners of the prismatoid. These lines determine for each edge of the prismatoid a plane triangle, and these triangles divide the prismatoid into the following parts:

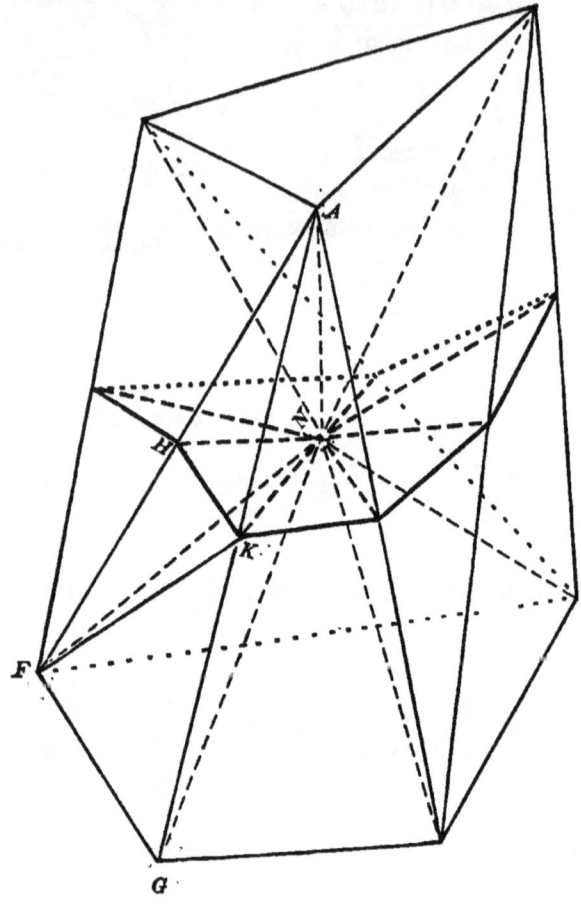

1. A pyramid whose vertex is N and whose base is B_2, the top of the prismatoid. Since the altitude of this pyramid is half that of the prismatoid, therefore, by 80, its volume is $\frac{1}{6} a B_2$.

2. A pyramid whose vertex is N and whose base is B_1, the bottom of the prismatoid. Since the altitude of this pyramid also is $\frac{1}{2} a$, therefore its volume is $\frac{1}{6} a B_1$.

THE MEASUREMENT OF VOLUMES. 103

3. Tetrahedra, like $ANFG$, each of which can have its volume expressed in terms of its own part of the midcross-section. For, let NH and NK be the lines in which the two sides ANF, ANG of the tetrahedron cut the mid-cross-section; and consider the part $ANHK$ of the tetrahedron $ANFG$. This part $ANHK$ is a pyramid whose base is the triangle NHK, and whose altitude is $\frac{1}{2}a$, half the altitude of the prismatoid. Hence, by 80, the volume of $ANHK$ is $\frac{1}{6}a(NHK)$. But, drawing KF, by 79,

and
$$ANHK = \tfrac{1}{2}ANFK,$$
$$ANFK = \tfrac{1}{2}ANFG.$$
Therefore,
$$ANFG = \tfrac{2}{3}a(NHK).$$

In like manner, the volume of every such tetrahedron is $\frac{2}{3}a$ times the area of its own piece of the midcross-section, and their sum is $\frac{2}{3}aM$. Now, combining 1, 2, and 3, which together make up the whole volume of the prismatoid, we find

$$D = \tfrac{1}{6}aB_2 + \tfrac{1}{6}aB_1 + \tfrac{2}{3}aM = \tfrac{1}{6}a(B_1 + 4M + B_2).$$

EXAM. 79. Given the plan of an embankment cut perpendicularly by the plane $AEID$, its top the pentagon $EFGHI$, its bottom the trapezoid $ABCD$, with the following measurements: For the lower base, $AB = 90$ meters, $CD = 110$ meters, $AD = 65$ meters; for the upper base $EF = 70$ meters, $EI = 30$ meters, $MF = MH = MG = 15$ meters; the breadth of the scarp $AE = 20$ meters, $DI = 15$ meters; the altitude of the embankment $a = 15$ meters. Find its volume.

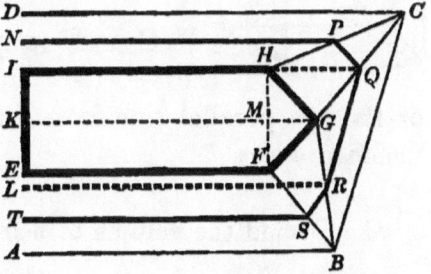

Here, for the midcross-section, we get

$$TS = 80 \text{ meters,}$$
$$LR = 87\cdot5 \text{ meters,}$$
$$IQ = 97\cdot5 \text{ meters,}$$
$$NP = 90 \text{ meters,}$$
$$TL = 7\cdot5 \text{ meters,}$$
$$LI = 32\cdot5 \text{ meters,}$$
$$IN = 7\cdot5 \text{ meters.}$$

Thus, the areas are

$$B_1 = 6500 \text{ square meters,}$$
$$B_2 = 2325 \text{ square meters,}$$
$$M = 4337\cdot5 \text{ square meters,}$$

and for the whole volume we get

$$D = 65,437\cdot5 \text{ cubic meters. } Ans.$$

NOTE. In a prismoid the midcross-section has always the same angles and the same number of sides as each base, every side being half the sum of the two corresponding basal edges. The rectangular prismoid has its top and bottom rectangles; hence, by 32, its volume

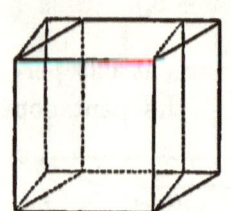

$$D = \tfrac{1}{6}a(R_1 + 4R + R_2)$$
$$= \tfrac{1}{6}a(w_1 b_1 + 4\frac{w_1 + w_2}{2} \times \frac{b_1 + b_2}{2} + w_2 b_2)$$
$$= \tfrac{1}{6}a(2w_1 b_1 + w_1 b_2 + w_2 b_1 + 2w_2 b_2).$$

Cor. If a prism has trapezoids for bases, its volume equals half the sum of its two parallel side-faces multiplied by their normal distance apart.

83. To find the volume of a frustum of a pyramid.

Rule: *To the areas of the two ends of the frustum add the square root of their product; multiply this sum by one-third the altitude.*

Formula: V. $F = \tfrac{1}{3}a(B_1 + \sqrt{B_1 B_2} + B_2).$

Proof: If w_1 and w_2 are two corresponding sides of the bases B_1 and B_2, then a side of the midsection is $\dfrac{w_1 + w_2}{2}$.

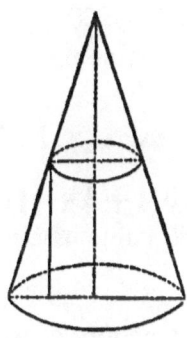 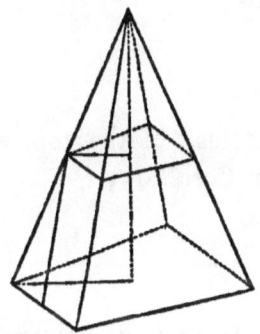

Since in a frustum B_1, B_2, and M are similar, by 44, we have

$$w_1 : \frac{w_1 + w_2}{2} :: \sqrt{B_1} : \sqrt{M},$$

and
$$w_2 : \frac{w_1 + w_2}{2} :: \sqrt{B_2} : \sqrt{M};$$

whence
$$w_1 + w_2 : \frac{w_1 + w_2}{2} :: \sqrt{B_1} + \sqrt{B_2} : \sqrt{M}.$$

Hence
$$2\sqrt{M} = \sqrt{B_1} + \sqrt{B_2},$$
and
$$4M = B_1 + 2\sqrt{B_1 B_2} + B_2.$$

Substituting this in 82 gives

$$\text{V. F} = \tfrac{1}{6} a(2B_1 + 2\sqrt{B_1 B_2} + 2B_2).$$

Cor. By 44,
$$B_2 = \frac{B_1 w_2^2}{w_1^2}.$$

Substituting this for B_2 gives

$$\text{V. F} = \tfrac{1}{3} a B_1 \left(1 + \frac{w_2}{w_1} + \frac{w_2^2}{w_1^2}\right).$$

106   MENSURATION.

EXAM. 80. The area of the top of a frustum is 160 square meters; of the bottom, 250 square meters; and its altitude is 24 meters. Find its volume.
Here
$$V. F = 8(250 + 200 + 160)$$
$$= 4880 \text{ cubic meters. } Ans.$$

If, instead of the top, we are given $w_1 : w_2 :: 5 : 4$, then, by our Corollary,
$$V. F = 2000(1 + \tfrac{4}{5} + \tfrac{16}{25})$$
$$= 4880 \text{ cubic meters. } Ans.$$

84. To find the volume of a frustum of any cone.

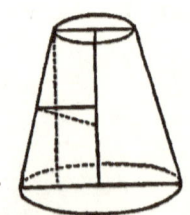

Rule: *To the areas of the two ends add the square root of their product; multiply this sum by one-third the altitude.*

Formula for Circular Cone:
$$V. F = \tfrac{1}{3} a \pi (r_1^2 + r_1 r_2 + r_2^2).$$

Proof: As in 81, so the frustum of a cone is the limit of the frustum of a pyramid.

EXAM. 81. The radius of one end is 5 meters; of the other, 3 meters; the altitude, 8 meters. Find the volume.
Here
$$V. F = \tfrac{1}{3} 8 (25 + 15 + 9) 3 \cdot 1416$$
$$= 410 \cdot 5024 \text{ cubic meters. } Ans.$$

85. To find the volume of any solid bounded terminally by two parallel planes, and laterally by a surface generated by the motion of a straight line always intersecting the planes, and returning finally to its initial position.

THE MEASUREMENT OF VOLUMES.

Rule: *Add the areas of the two ends to four times the midsection; multiply this sum by one-sixth the altitude.*

Prismoidal Formula: $D = \tfrac{1}{6} a (B_1 + 4M + B_2)$.

Proof: Join neighboring points in the top perimeter of such a solid to form a polygon, likewise in the perimeter of the bottom. Take the two polygons so formed as bases

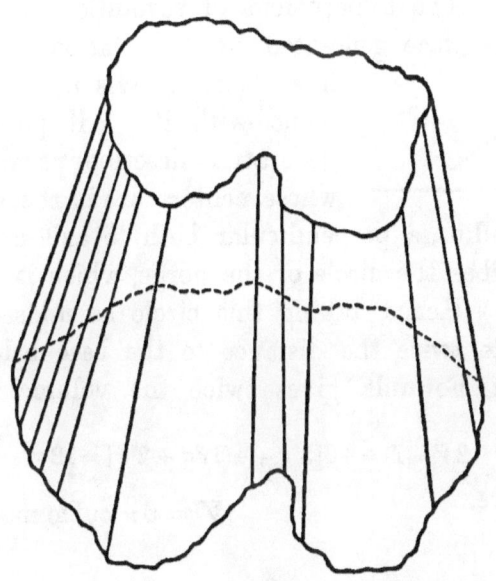

of a prismatoid. Then when the number of basal edges is indefinitely increased, each edge decreasing indefinitely in length, as thus its bases approach to coincidence with the bases of the solid, the sides of the prismatoid approach the ruled surface, and its volume and midsection approach the volume and midsection of the solid as limit. But always the variable volume is to the variable sum $(B_1 + 4M + B_2)$ in the constant ratio $\tfrac{1}{6} a$. Therefore, by

13, their limits will be to one another in the same ratio; and
$$D = \tfrac{1}{6}a(B_1 + 4M + B_2)$$
for the prismoidal solid.

EXAM. 82. The radius of the minimum circle in a hyperboloid is 1 meter. Find the volume contained between this *circle of the gorge* and a circle 3 meters below it whose radius is 2 meters.

Solution: The hyperboloid of revolution of one nappe is a ruled surface generated by the rotation of a straight line about an axis not in the same plane with it. All points of the generatrix describe parallel circles whose centers are in the axis. The shortest radius, a perpendicular both to axis and generatrix, describes the circle of the gorge, which is a plane of symmetry. Hence, taking this circle as midsection, and for altitude twice the distance to the base below it, the Prismoidal Formula gives twice the volume sought in Exam. 82.

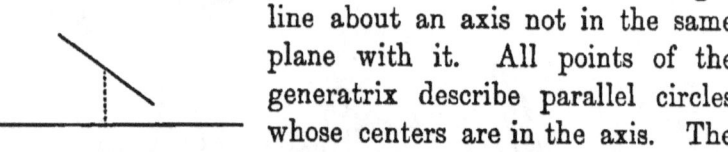

$$2V = D = \tfrac{1}{6}6[2^2\pi + 4(1)^2\pi + 2^2\pi] = 12\pi.$$

Therefore, $V = 6\pi$ cubic meters. *Ans.*

86. To find the volume of any wedge.

Rule: *To twice the length of the base add the opposite edge; multiply the sum by the width of the base, and this product by one-sixth the altitude of the wedge.*

Formula: $W = \tfrac{1}{6}aw(2b_1 + b_2).$

Proof: Since the upper base of a wedge is a line, so, by the Prismoidal Formula,
$$W = \tfrac{1}{6}a(B_1 + 4M).$$

THE MEASUREMENT OF VOLUMES.

Therefore, by 32,

$$W = \tfrac{1}{6} a \left(w b_1 + 4 \frac{b_1 + b_2}{2} \times \frac{w}{2} \right) = \tfrac{1}{6} a w (b_1 + b_1 + b_2).$$

Cor. 1. If the length of edge equals the length of base; *i.e.*,
$$b_1 = b_2, \quad \text{then} \quad W = \tfrac{1}{2} a w b,$$
the simplest form of wedge.

Cor. 2. The volume of any truncated triangular prism is equal to the product of its right section by one-third the sum of its lateral edges.

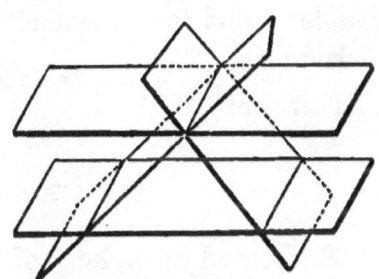

EXAM. 83. Find the volume of a wedge, of which the length of the base is 70 meters; the width, 30 meters; the length of the edge, 110 meters; and the altitude, 24·8 meters.

Here
$$\begin{aligned} W &= (140 + 110) \tfrac{30 \times 24 \cdot 8}{6} \\ &= (140 + 110) 10 \times 12 \cdot 4 \\ &= 2500 \times 12 \cdot 4 \\ &= 31{,}000 \text{ cubic meters. } Ans. \end{aligned}$$

87. To find the volume of any tetrahedron.

Rule : *Multiply double the area of a parallelogram whose vertices bisect any four edges by one-third the perpendicular to both the other edges.*

Formula : $X = \tfrac{2}{3} a M.$

Proof : When, the bases being lines,
$$B_1 = B_2 = 0, \quad \text{then} \quad D = X = \tfrac{1}{6} a 4 M = \tfrac{2}{3} a M.$$

110   MENSURATION.

Since M bisects the line perpendicular to the two basal edges, it bisects the four lateral edges; and is a parallelogram.

EXAM. 84. The line perpendicular to both basal edges of a tetrahedron is $2r$ long; the length of the top edge is $2r$, and of the bottom edge, $2\pi r$. The midsection is a rectangle. Find the volume of the tetrahedron.
Here
$$a = 2r, \text{ and } M = r^2\pi.$$
Therefore, $X = \frac{4}{3}\pi r^3.$ *Ans.*

§ (P). SPHERE.

88. To find the volume of a sphere.

Rule: *Multiply the cube of its radius by* 4·1888−.

Formula: V. H $= \frac{4}{3}\pi r^3$.

Proof: Any sphere is equal in volume to a tetrahedron whose midsection is equivalent to a great circle of the sphere, and whose altitude equals a diameter.

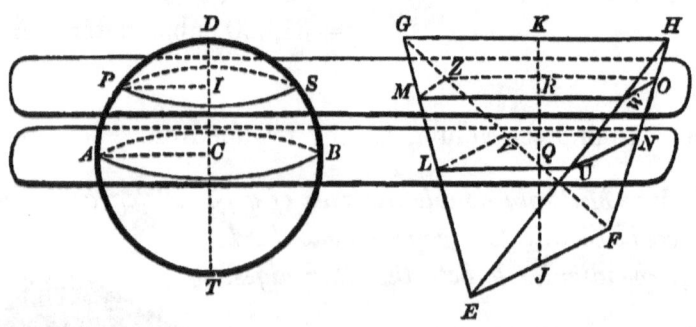

Let the diameter DC be normal to the great circle AB at C. Let Q be the point in which the midsection LN bisects the altitude JK at right angles. In both solids take any height $CI = QR$, and through the points I and R the sections PS parallel to AB, and MO parallel to LN. Then, in the sphere, by 47,

$$\odot AB : \odot PS :: AC^2 : PI^2,$$

or, by Ww. 289; (Eu. VI. 8, Cor.; Cv. III. 44),

$$\odot AB : \odot PS :: AC^2 : TI \times ID \quad \ldots \ldots \quad (1).$$

In the tetrahedron, by Ww. 279 & 469; (Eu. VI. 4, & XI. 17; Cv. IV. 25, & VI. 37),

$$LU : MW :: EL : EM :: JQ : JR;$$
$$LV : MZ :: GL : GM :: KQ : KR;$$

and since, by Ww. 315; (Eu. VI. 23; Cv. IV. 5),

$$\square LN : \square MO :: LU \times LV : MW \times MZ,$$
therefore, $\square LN : \square MO :: JQ \times QK : JR \times RK \quad . \quad . \quad (2).$

But now, by hypothesis and construction, in proportions (1) and (2), the first, third, and fourth terms are respectively equal, therefore

$$\odot PS = \square MO;$$

and since these are corresponding sections at any height, therefore, by Scholium to 79, the sphere and tetrahedron are equal in volume.

Thus, by 87,

$$\text{V. H} = X = \tfrac{2}{3}aM = \tfrac{2}{3} 2rr^2\pi = \tfrac{4}{3}\pi r^3.$$

EXAM. 85. If, in making a model of the tetrahedron $EFGH$, we wish the midsection LN to be a square, and the four lateral edges equal, find in terms of radius their length and that of the two basal edges.

By hypothesis,

square $LN = r^2\pi; \quad \therefore \ LU = r\sqrt{\pi}.$

But $\quad GH = 2LU = 2r\sqrt{\pi}$,
and $\quad EF = 2LV = 2LU = 2r\sqrt{\pi}$.

For any one of the equal lateral edges,

$\overline{EG}^2 = \overline{GK}^2 + \overline{KE}^2 = \overline{GK}^2 + \overline{KJ}^2 + \overline{JE}^2 = \overline{LU}^2 + \overline{DT}^2 + \overline{LV}^2$.

$\therefore EG = \sqrt{r^2\pi + 4r^2 + r^2\pi} = \sqrt{4r^2\left(1 + \frac{\pi}{2}\right)} = 2r\sqrt{1 + \frac{\pi}{2}}$.

So it is not a regular tetrahedron.

89. To find the volume of any spherical segment.

Rule: *To three times the sum of the squared radii of the two ends add the squared altitude; multiply this sum by the altitude, and the product by ·5236−.*

Formula: $V. G = \tfrac{1}{6} a\pi [3(r_1^2 + r_2^2) + a^2]$.

Proof: In 88, we proved any spherical segment equal in volume to a prismoid of equivalent bases and altitude. Therefore, by 82,

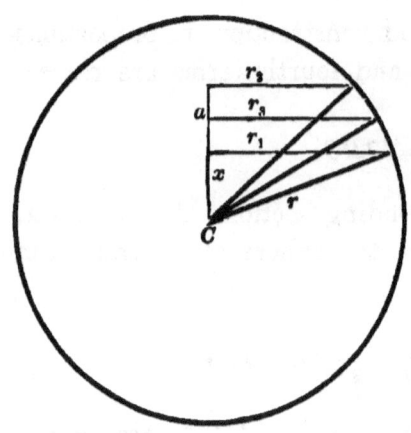

$V. G = \tfrac{1}{6} a(r_1^2\pi + 4r_3^2\pi + r_2^2\pi)$.

To eliminate r_3 call x the distance from center of sphere to bottom of segment, and r the radius of sphere; then, by Eu. II. 10,

$(a+x)^2 + x^2 = 2\left(\frac{a}{2}\right)^2 + 2\left(\frac{a}{2}+x\right)^2$.

Doubling and subtracting both members from $4r^2$, gives

$2r^2 - 2(a+x)^2 + 2r^2 - 2x^2 = 4r^2 - 4\left(\frac{a}{2}+x\right)^2 - a^2$,

or, by 2, $\quad 2r_2^2 + 2r_1^2 + a^2 = 4r_3^2$.

Substituting, $\quad V. G = \tfrac{1}{6} a\pi (3r_1^2 + 3r_2^2 + a^2)$.

Cor. In a segment of one base, since $r_2 = 0$, we have
$$V. G = \tfrac{1}{6} a \pi (3 r_1^2 + a^2).$$
But now, by Ww. 289; (Eu. VI. 8, Cor.; Cv. III. 44),
$$r_1^2 = a(2r - a).$$
Substituting,
$$V. G = \tfrac{1}{6} a \pi [3 a (2r - a) + a^2] = \tfrac{1}{6} a \pi (6 a r - 3 a^2 + a^2) = a^2 \pi (r - \tfrac{1}{3} a).$$

EXAM. 86. If the axis of a cylinder passes through the center of a sphere, the sphere-ring so formed is equal in volume to a sphere of the same altitude.

For, since the bases of a middle segment are equidistant from the center,
$$V. G = \tfrac{1}{6} a \pi (6 r_1^2 + a^2)$$
$$= a \pi r_1^2 + \tfrac{1}{6} \pi a^3.$$

But, by 76, the volume of the cylinder cut out of the segment is $a r_1^2 \pi$, and the remaining ring $\tfrac{1}{6} \pi a^3$ is, by 88, the volume of a sphere of diameter a.

XLII. When a semicircle revolves about its diameter, the solid generated by any sector of the semicircle is called a *spherical sector*.

90. To find the volume of any spherical sector.

Rule: *Multiply its zone by one-third the radius.*

Formula: $V. S = \tfrac{2}{3} \pi a r^2.$

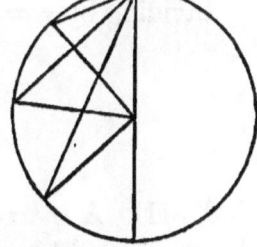

Proof: If one radius of the generating sector coincides with the axis of revolution, the spherical sector is the sum of a spherical segment of one base and a cone on same base, with vertex at center of sphere.

By 89, *Cor.*,   V. G $= a^2\pi(r - \tfrac{1}{3}a)$.
By 81,   V. K $= \tfrac{1}{3}(r-a)r_1^2\pi$;

or, substituting for r_1^2, its value used in 89, *Cor.*,

V. K $= \tfrac{1}{3}(r-a)a(2r-a)\pi = \tfrac{1}{3}\pi(2r^2a - 3ra^2 + a^3)$.

Adding, we have

V. S $=$ V. G $+$ V. K $= \tfrac{2}{3}a\pi r^2$.

Any other spherical sector is the difference of two such sectors.

V. S $=$ V. S$_1$ $-$ V. S$_2$ $= \tfrac{2}{3}r^2\pi a_1 - \tfrac{2}{3}r^2\pi a_2 = \tfrac{2}{3}r^2\pi(a_1 - a_2)$.

But $a_1 - a_2 = a$,

the altitude of S's zone, whose area, by 65, is $2\pi ra$. Thus, for every spherical sector the volume is *zone by* $\tfrac{1}{3}r$.

Cor. If r_1 and r_2 are the radii of the bases of the zone, its altitude,

$$a = \sqrt{r^2 - r_2^2} - \sqrt{r^2 - r_1^2}.$$

EXAM. 87. Find the diameter of a sphere of which a sector contains 7·854 cubic meters when its zone is 0·6 meters high.

V. S $= \tfrac{2}{3} 0·6\,\pi r^2 = 0·4\,\pi r^2 = 7·854$.

Dividing by $\pi = 3·1416$,

$0·4\,r^2 = 2·5$. $\therefore\ 4r^2 = 25$.

$\therefore\ 2r = 5$ meters. *Ans.*

XLIII. A *spherical ungula* is a part of a globe bounded by a lune and two great semicircles.

91. To find the volume of a spherical ungula.

Rule: *Multiply the area of its lune by one-third the radius.*

Formula: $\hat{v} = \tfrac{1}{3} r^3 u$.

Proof:
By 69, $\quad \hat{v} : V . H :: L : H$.

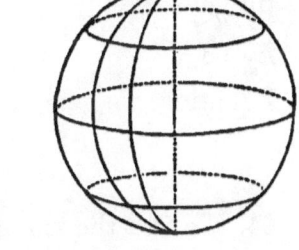

$\therefore \hat{v} : \tfrac{4}{3}\pi r^3 :: 2r^2 u : 4r^2\pi$.

$\therefore \hat{v} = \tfrac{1}{3} r^3 u$.

Cor. On equal spheres, ungulae are as their angles.

92. To find the volume of a spherical pyramid.

Rule: *Multiply the area of its base by one-third the radius.*

Formula: $\widehat{Y} = \tfrac{1}{3} r^3 e$.

Proof:
By 69, $\quad \widehat{Y} : V . H :: \widehat{N} : H$.

$\therefore \widehat{Y} : \tfrac{4}{3} r^3\pi :: er^2 : 4r^2\pi$.

$\therefore Y = \tfrac{1}{3} r^3 e$.

₴ (Q). THEOREM OF PAPPUS.

93. If a plane figure, lying wholly on the same side of a line in its own plane, revolves about that line, the volume of the solid thus generated is equal to the product of the revolving area by the length of the path described by its center of mass.

Scholium. As for 66, so we give under 94, by a single representative case, the general demonstration for all figures having an axis of symmetry parallel to the axis of revolution.

116   MENSURATION.

Exam. 88. Find the distance of the center of mass of a semicircle from the center of the circle.

By 88,   V. II $= \tfrac{2}{3}\pi r^3$.
By 93,   V. H $= \tfrac{1}{2}r^2\pi \cdot 2x\pi$.
Equating, we get $\tfrac{2}{3}\pi r^3 = r^2\pi^2 x$.
       $\therefore \tfrac{2}{3}r = \pi x$.

$$\therefore x = \frac{4r}{3\pi}. \text{ Ans.}$$

94. To find the volume of a ring.

Rule : *Multiply the generating area by the path of its center.*

Formula for Ellipse : V. O $= 2\pi^2 abr$.

Proof : Conceive any ellipse to revolve about an exterior axis parallel to one of its axes. Divide the axis of symmetry AE into n equal parts, as

$$AV = VT = z,$$

and from these points of division drop perpendiculars on the axis of revolution PO. Join the points where these perpendiculars cut the ellipse by chords FD, DA, AG, GH, etc.

The volume generated by one of the trapezoids thus formed, as $DGHF$, is the difference between the frustums generated by the right-angled trapezoids $QDFW$ and $QGHW$. Therefore, by 84,

V. by $DGHF$
  $= \tfrac{1}{3}z\pi(FW^2 + FW, DQ + DQ^2) - \tfrac{1}{3}z\pi(HW^2 + HW, GQ + GQ^2)$
  $= \tfrac{1}{3}z\pi[(CO+FT)^2 + (CO+FT)(CO+DV) + (CO+DV)^2$
    $- (CO-FT)^2 - (CO-FT)(CO-DV) - (CO-DV)^2]$
  $= \tfrac{1}{3}z\pi(6CO, FT + 6CO, DV)$
  $= 2\pi, CO, z(FT+DV)$
  $= 2\pi r z \dfrac{FH + DG}{2}$.

Thus, by 40, the volume generated by the polygon *HGADF*, etc., equals its area multiplied by the path of the center. But, as we increase *n*, and thus decrease *z* indefinitely, as shown in 55, the area of the polygon ap-

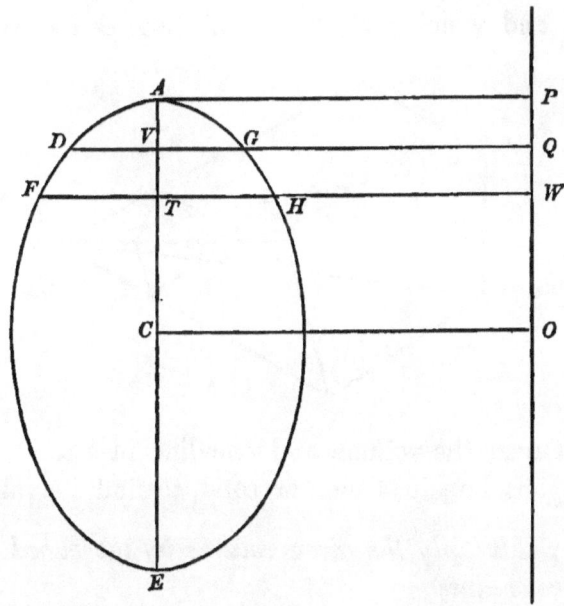

proaches the area of the ellipse as its limit. But always the variable volume is to the variable area in the constant ratio $2\pi r$; therefore, by 13, their limits will be to one another in the same ratio; and

$$V. O = 2\pi r ab\pi.$$

EXAM. 89. Find the volume of the ring swept out by an ellipse whose axes are 8 and 16 meters, revolving round an axis in its own plane, and 10 meters from its center.
   Here
$$V. O = 4 \times 8 \times 10 \times 2\pi^2$$
$$= 640\pi^2$$
$$= 6316 \cdot 5 \text{ cubic meters. } Ans.$$

§ (R). SIMILAR SOLIDS.

XLIV. *Similar polyhedrons* are those bounded by the same number of faces respectively similar and similarly placed, and which have their solid angles congruent.

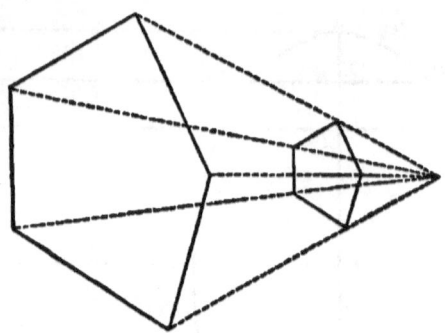

95. Given the volume and one line in a solid, and the homologous line in a similar solid, to find its volume.

Rule: *Multiply the given volume by the cubed ratio of homologous lines.*

Formula: $V_1 = \dfrac{V_2 a_1^3}{a_2^3}$.

Proof: The volumes of two similar solids are as the cubes of any two corresponding dimensions.

Ww. 590; (Eu. XI. 33; Cv. VII. 73).

Thus, for the sphere, by 88,

$$\frac{H_1}{H_2} = \frac{\frac{4}{3}\pi r_1^3}{\frac{4}{3}\pi r_2^3} = \frac{r_1^3}{r_2^3}.$$

NOTE. If a tetrahedron is cut by a plane parallel to one of its faces, the tetrahedron cut off is similar to the first. If a cone be cut by a plane parallel to its base, the whole cone and the cone cut off are similar.

EXAM. 90. The edge of a cube is 1 meter; find the edge of a cube of double the volume.

The cube of the required number is to the cube of 1 as 2 is to 1; or,
$$x^3 : 1 :: 2 : 1.$$
$$\therefore x^3 = 2.$$
$$\therefore x = \sqrt[3]{2} = 1{\cdot}25992+. \ Ans.$$

Thus a cube, with its edge 1·26 meters, is more than double a cube with edge 1 meter.

EXAM. 91. The three edges of a quader are as 3, 4, 7, and the volume is 777,924; find the edges.

By 72, the volume of the quader, whose edges are 3, 4, 7, is 84; then
$$84 : 777{,}924 :: 3^3 : 250{,}047,$$

and
$$\sqrt[3]{250{,}047} = 63;$$
$$3 : 4 :: 63 : 84,$$
$$3 : 7 :: 63 : 127.$$

Therefore, the edges are 63, 84, 127. *Ans.*

§(8). IRREGULAR SOLIDS.

96. Any small solid may be estimated by placing it in a vessel of convenient shape, such as a quader or a cylinder, and pouring in a liquid until the solid is quite covered; then noting the level, removing the solid, and again noting the level at which the liquid stands. The volume of the solid is equal to the volume of the vessel between the two levels.

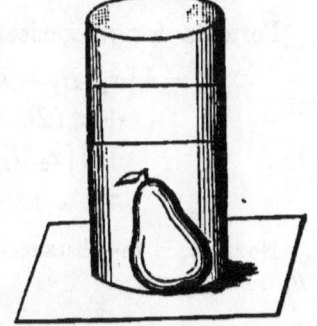

97. If the solid is homogeneous, weigh it. Also weigh a cubic centimeter of the same substance. Divide the weight of the solid by the weight of the cubic centimeter. The quotient will be the number of cubic centimeters in the solid.

From 73, we have the

Formula: $V^{ccm} = \dfrac{\omega^g}{\delta}$.

EXAM. 92. A ball 5 centimeters in diameter weighs 431·97 grams. An irregular solid of the same substance weighs 13·2 grams; find its volume.

The volume of the ball is

$$5^3 \times 0.5236 = 65.45.$$
$$\therefore 431.97 \div 65.45 = 6.6 \text{ grams,}$$

the weight of a cubic centimeter.

$$\therefore 13.2 \div 6.6 = 2 \text{ cubic centimeters. } Ans.$$

98. To find the volume of any irregular polyhedron.

Rule: *Cut the polyhedron into prismatoids by passing parallel planes through all its summits.*

Formula for n consecutive prismatoids:

$$I = \tfrac{1}{6}[x_2(B_1 - B_3) + x_3(B_2 - B_4) + \text{etc.}$$
$$+ x_n(B_{n-1} - B_{n+1}) + x_{n+1}(B_n + B_{n+1})]$$
$$+ \tfrac{2}{3}[x_2 M_1 + (x_3 - x_2)M_2 + (x_4 - x_3)M_3 + \text{etc.}$$
$$+ (x_{n+1} - x_n)M_n].$$

NOTE. x_2 is the distance of B_2 from B_1, and x_3 is the distance of B_3 from B_1, etc.

Proof: This formula is obtained directly by the method of 41.

CHAPTER VI.

THE APPLICABILITY OF THE PRISMOIDAL FORMULA.

99. To find whether the volume of any solid is determined by the Prismoidal Formula.

Rule: *The Prismoidal Formula applies exactly to* ALL SOLIDS *contained between two parallel planes,* OF WHICH *the area of any section parallel to these planes can be expressed by a rational integral algebraic function, of a degree not higher than the third, of its distance from either of these bounding planes or bases.*

Test: $A_x = q + mx + nx^2 + fx^3$.

NOTE. A_x is the area of any section of the solid at the distance x from one of its ends. The coefficients q, m, n, f, are constant for the same solid, but may be either positive or negative; or any one, two, or three of them may be zero.

Proof: Measuring x on a line normal to which the sections are made, let $\phi(x)$ be the area of the section at the distance x from the origin.

The problem then is, What function ϕ will fulfil the conditions of the Prismoidal Formula?

For any linear unit, the segment between $\phi(0)$ and $\phi(4)$ is the sum of the segments between $\phi(0)$ and $\phi(2)$ and between $\phi(2)$ and $\phi(4)$. Therefore, if ϕ is such a function

MENSURATION.

as to fulfil the requirements of the Prismoidal Formula, we have identically

$$\tfrac{4}{3}[\phi(0)+4\phi(2)+\phi(4)] = \tfrac{2}{3}[\phi(0)+4\phi(1)+\phi(2)] + \tfrac{2}{3}[\phi(2)+4\phi(3)+\phi(4)].$$

$$\therefore \phi(0) - 4\phi(1) + 6\phi(2) - 4\phi(3) + \phi(4) = 0.$$

But for $\phi(x) = q + mx + nx^2 + fx^3 + gx^4$,

$\phi(0) - 4\phi(1) + 6\phi(2) - 4\phi(3) + \phi(4)$

becomes

$$\begin{array}{r}
+\,q \\
-\,4q - 4m - 4n - 4f - 4g \\
+\,6q + 12m + 24n + 48f + 96g \\
-\,4q - 12m - 36n - 108f - 324g \\
+\,q + 4m + 16n + 64f + 256g \\ \hline
0 \quad 0 \quad 0 \quad 0 + 24g
\end{array}$$

So the conditions are satisfied only by functions which have no fourth and higher powers. Hence $\phi(x)$ must be an algebraic expression of positive integral powers not exceeding the third degree.

Thus, in general, the cubic equation

$$A_x = q + mx + nx^2 + fx^3$$

expresses the law of variation in magnitude of the plane generatrix of prismoidal spaces; *i.e.*, solids to which the Prismoidal Formula universally applies.

Cor. 1. Since for prismoidal solids

$$\phi(x) = n_0 + n_1 x + n_2 x^2 + n_3 x^3,$$

therefore, $\phi(0) + 4\phi(\tfrac{1}{2}a) + \phi(a) =$

$$\begin{array}{l}
n_0 \\
+\,4n_0 + 2an_1 + a^2 n_2 + \tfrac{1}{2} a^3 n_3 \\
+\,n_0 + an_1 + a^2 n_2 + a^3 n_3 \\ \hline
= 6n_0 + 3an_1 + 2a^2 n_2 + \tfrac{3}{2} a^3 n_3
\end{array}$$

THE APPLICABILITY OF THE PRISMOIDAL FORMULA. 123

Thus, $\quad D = \tfrac{1}{6} a (B_1 + 4M + B_2)$
$\quad\quad\quad = \tfrac{1}{6} a [\phi(0) + 4\phi(\tfrac{1}{2}a) + \phi(a)]$
$\quad\quad\quad = a n_0 + \tfrac{1}{2} a^2 n_1 + \tfrac{1}{3} a^3 n_2 + \tfrac{1}{4} a^4 n_3.$

Cor. 2. Of any solid whose

$$A_x = \phi(x) = n_0 + n_1 x + n_2 x^2 + n_3 x^3 + n_4 x^4 + \cdots + n_m x^m,$$

the volume is

$$a n_0 + \tfrac{1}{2} a^2 n_1 + \tfrac{1}{3} a^3 n_2 + \tfrac{1}{4} a^4 n_3 + \tfrac{1}{5} a^5 n_4 + \cdots + \frac{1}{m+1} a^{m+1} n_m.$$

For the volume of the prism whose base is the cross-section $\phi(x)$, and whose altitude is the nth part of the altitude of the whole solid, is $\dfrac{a}{n} \phi(x)$.

The limit of the sum of all the prisms of like height

$$\frac{a}{n}\left[\phi(0) + \phi\left(\frac{1}{n}a\right) + \phi\left(\frac{2}{n}a\right) + \cdots + \phi\left(\frac{n-1}{n}a\right)\right],$$

when n becomes indefinitely great, is the volume of the whole solid.

But $\quad \dfrac{1^r + 2^r + \cdots + (n-1)^r}{n^{r+1}} = \dfrac{1}{r+1},\quad$ when $n \doteq \infty$.

§ (T). PRISMOIDAL SOLIDS OF REVOLUTION.

The general expression

$$A_x = q + mx + nx^2 + fx^3,$$

has as many possible varieties as there are combinations of four things taken one, two, three, and four together; that is, $2^4 - 1$, or 15 varieties.

Corresponding to each of these there will be *at least one solid of revolution* generated by the curve whose equation is, in the general case,

$$\pi y^2 = q + mx + nx^2 + fx^3.$$

For, if y be the revolving ordinate of any point in the curve, then πy^2 is the area of the section at distance x from one end of the solid.

XLV. Examination of the Different Cases.

(1) Let $\pi y^2 = q$; $\therefore y$ is constant, and the solid is a circular cylinder.

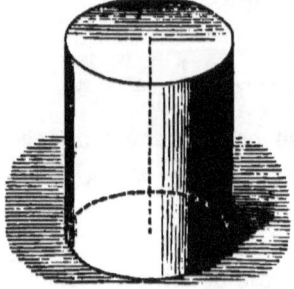

(2) Let $\pi y^2 = mx$; $\therefore y^2 \propto x$, and the solid is a paraboloid of revolution; for, in a parabola, the square of the ordinate varies as the abscissa.

(3) Let $\pi y^2 = nx^2$; $\therefore y \propto x$, and the solid is a right circular cone.

(4) Let $\pi y^2 = fx^3$; $\therefore y^2 \propto x^3$, and the solid is a semicubic paraboloid of revolution.

(5) Let $\pi y^2 = q + mx$; $\therefore y^2 \propto (h + x)$ where h is constant, and the solid is a *frustum* of a paraboloid of revolution, h being the height of the segment cut off.

(6) Let $\pi y^2 = q + nx^2$; supposing q and n positive, this is the equation to a hyperbola, the conjugate axis being the axis of x, and the center the origin. Hence, the solid is a hyperboloid of one nappe.

(7) Let $\pi y^2 = q + fx^3$. In this case, the solid is generated by the revolution of a curve, somewhat similar in form to the semicubic parabola, round a line parallel to the axis of x, and at a constant distance from it.

(8) Let $\pi y^2 = mx + nx^2$. In this case, the solid may be a sphere, a prolate spheroid, an oblate spheroid, a hyperboloid of revolution, or its conjugate hyperboloid.

(9) Let $\pi y^2 = q + mx + nx^2$. In this case, the solid will be a *frustum* of a circular cone, or of the sphere, spheroids, or hyperboloids of revolution, made by planes normal to the axis. In the frustum of the cone q, m, and n are all positive. The other solids in (8) and (9) are distinguished by the values and signs of the constants m and n.

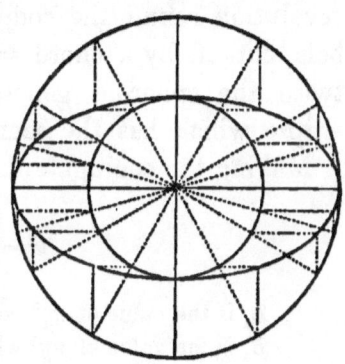

(10) Let $\pi y^2 = q + mx + nx^2 + fx^3$. In this case, the solid is a frustum of a semicubic paraboloid of revolution. For, if x be the distance of the section from the smaller end of the frustum, and h the height of the segment cut off, $\therefore y^2 \propto (h+x)^3$. $\therefore \pi y^2$ is *of the form* $q + mx + nx^2 + fx^3$.

(11) Let $\pi y^2 = mx + fx^3$.
(12) Let $\pi y^2 = nx^2 + fx^3$.
(13) Let $\pi y^2 = q + mx + fx^3$.
(14) Let $\pi y^2 = q + nx^2 + fx^3$.
(15) Let $\pi y^2 = mx + nx^2 + fx^3$.

EXAM. 93. Since, for an oblate spheroid,

$$B_1 = 0, \quad B_2 = 0, \quad 4M = 4\pi a^2, \quad \text{and} \quad h = 2b,$$

therefore its volume

$$\sigma = \tfrac{4}{3} \pi a^2 b.$$

Similarly, the volume of a prolate spheroid

$$\theta = \tfrac{4}{3} \pi a b^2.$$

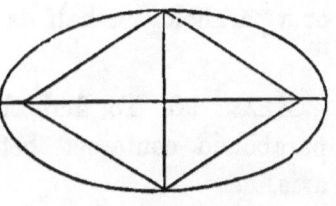

Thus each is, like the sphere, two-thirds of the circumscribed cylinder.

EXAM. 94. The volume of the solid generated by the revolution round the conjugate axis of an arc of a hyperbola, cut off by a chord = and ‖ to the conjugate axis, is twice the spheroid generated by the revolution of the ellipse which has the same axes.

Making the conjugate the axis of x,

$$y^2 = \frac{a^2}{b^2}(b^2 + x^2).$$

B_1 is the value of πy^2 when $x = b$; $\therefore B_1 = 2\pi a^2$.
B_2 is the value of πy^2 when $x = -b$; $\therefore B_2 = 2\pi a^2$.
M is the value of πy^2 when $x = 0$; $\therefore 4M = 4\pi a^2$.

Since the conjugate axis is the height of the solid,

$$\therefore h = 2b.$$

Hence its volume $\chi = \tfrac{8}{3}\pi a^2 b$. *Ans.*

EXAM. 95. To find the volume of a paraboloid of revolution. Let h be its height, that is, the length of the axis, r the radius of its base, and p the parameter of the generating parabola, $y^2 = px$.

Then
$$B_1 = 0, \quad B_2 = \pi r^2 = \pi p h, \quad 4M = 4\pi p\frac{h}{2} = 2\pi p h.$$

$$\therefore \zeta = \tfrac{1}{2}\pi p h^2. \ \textit{Ans.}$$

Cor. Since $r^2 = ph$, $\therefore \zeta = \tfrac{1}{2}\pi p h h = \tfrac{1}{2}\pi r^2 h$,

or a paraboloid is half its circumscribing cylinder.

EXAM. 96. To find the volume of any frustum of a paraboloid contained between two planes normal to the axis.

F. $\zeta = \tfrac{1}{2}\pi p h_2^2 - \tfrac{1}{2}\pi p h_1^2 = \tfrac{1}{2}\pi p(h_2^2 - h_1^2) = \tfrac{1}{2}\pi p(h_2 + h_1)(h_2 - h_1).$

But $r_1^2 = ph_1$,
and $r_2^2 = ph_2$,
$$\therefore h_1 + h_2 = \frac{1}{p}(r_1^2 + r_2^2);$$
and $h_2 - h_1 = a$,

the altitude of the frustum.
$$\therefore F. \zeta = \tfrac{1}{2}\pi p \frac{1}{p}(r_1^2 + r_2^2)a$$
$$= \tfrac{1}{2}\pi a(r_1^2 + r_2^2). \ Ans.$$

§ (U).—**PRISMOIDAL SOLIDS NOT OF REVOLUTION.**

We may now consider the same fifteen possible varieties when A_x is not of the form πy^2.

XLVI. Discussion of Cases.

(1) Let $A_x = q$. In this case all the transverse sections are constant. This is the property of all prisms and cylinders; also, of all solids uniformly twisted, *e.g.*, the square-threaded screw.

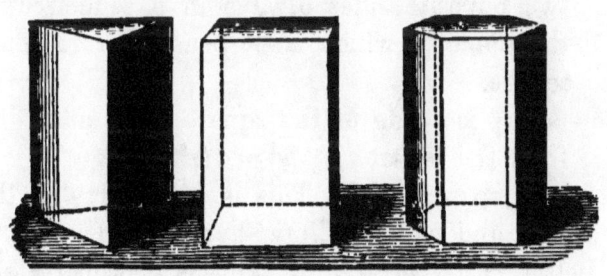

(2) Let $A_x = mx$. This is a property of the elliptic paraboloid, or the solid generated by the motion of a variable ellipse whose axes are the double ordinates of two parabolas which have a common axis and a

common vertex, the plane of the ellipse being always normal to this axis; for, in this solid, the area of the section at distance x from the vertex will be $\pi y y'$, where y and y' are the ordinates of the two parabolas, and since both y^2 and y'^2 vary as x, $\therefore \pi y y'$ varies as x.

(3) Let $A_x = nx^2$. This is a property of all pyramids and cones, whatever may be their bases.

(4) Let $A_x = fx^3$. The solid will be an elliptic semicubic paraboloid. Substitute semicubic for common parabolas in (2).

(5) Let $A_x = q + mx$. This is a property of a *frustum* of an elliptic paraboloid.

(6) Let $A_x = q + nx^2$. This is a property of a *groin*, a simple case of which is the square groin seen in the vaults of large buildings.

This solid may be generated by a variable square, which moves parallel to itself, with the midpoints of two opposite sides always in a semicircumference, the plane of which is perpendicular to that of the square.

If y is a side of the square when at a distance x from the centre; $\therefore y^2 = 4(r^2 - x^2)$.

(7) Let $A_x = mx + nx^2$. This is a property of the ellipsoid, and of the elliptic hyperboloid.

(8) Let $A_x = q + mx + nx^2$. This is a property of a prismoid, and of any frustum of a pyramid or cone, whatever may be the base; also, of any frustum of an ellipsoid, or elliptic hyperboloid made by planes perpendicular to the axis of the hyperboloid, or to any one of the three axes of the ellipsoid.

THE APPLICABILITY OF THE PRISMOIDAL FORMULA. 129

Exam. 97. To find the volume of an ellipsoid. Let a, b, c be the three semi-axes, a the greatest;

$$\therefore h = 2a, \quad B_1 = 0, \quad B_2 = 0, \quad 4M = 4\pi bc.$$

$$\therefore \eta = \tfrac{4}{3}\pi abc. \quad Ans.$$

§ (V). ELIMINATION OF ONE BASE.

For all solids whose section is a function of degree not higher than the second, or

$$A_x = q + mx + nx^2,$$

q, m, n, and consequently A_x, for all values of x, are determined if the value of A_x for three values of x is known.

Measuring x from B_1 we have

$$A_0 = B_1 = q.$$

Supposing we know the section at $\dfrac{1}{z}$ the height of the solid above B_1, we have for determining m and n the two equations,

$$B_2 = B_1 + ma + na^2,$$
$$A_{\frac{a}{z}} = B_1 + m\frac{a}{z} + n\frac{a^2}{z^2}.$$

Hence,

$$m = \frac{z^2 A_{\frac{a}{z}} - (z^2 - 1)B_1 - B_2}{(z-1)a},$$

$$n = \frac{zB_2 + z(z-1)B_1 - z^2 A_{\frac{a}{z}}}{(z-1)a^2}.$$

For the volume of the solid we have

$$V = B_1 a + \tfrac{1}{2}ma^2 + \tfrac{1}{3}na^3,$$

or

$$V = \frac{a}{6(z-1)}[(2z-3)B_2 - (z-1)(z-3)B_1 + z^2 A_{\frac{a}{z}}].$$

For $z = 3$, this gives
$$V = \frac{a}{4}(B_2 + 3A_{\frac{2}{3}}).$$
Again, for $z = 1\frac{1}{2}$,
$$V = \frac{a}{4}(B_1 + 3A_{\frac{1}{3}a}).$$

These give the following theorem:

100. To find the volume of a prismatoid, or of any solid whose section gives a quadratic:

Rule: *Multiply one fourth its altitude by the sum of one base and three times a section distant from that base two-thirds the altitude.*

Cor. If B_2 reduces to an edge or a point,
$$V = \tfrac{3}{4} a A_{\frac{2}{3}}.$$

CHAPTER VII.

APPROXIMATION TO ALL SURFACES AND SOLIDS.

§ (W). WEDDLE'S METHOD.

101. To find the content between the first and seventh of equidistant sections:

Weddle's Rule: *To five times the sum of the even sections add the middle section, and all the odd sections; multiply this sum by three-tenths of the common distance between the sections.*

Formula: $\xi = \tfrac{3}{10} h [5(y_2 + y_4 + y_6) + y_4 + y_1 + y_3 + y_5 + y_7]$.

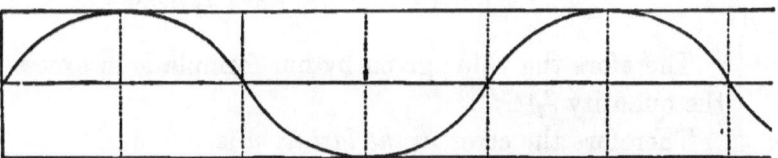

Proof: If we take the origin midway between the ends (of the solid or plane figure), and suppose every section y perpendicular to the axis expressible as an algebraic function of positive integer powers of x not exceeding the seventh degree, then for every

$$y = \phi(x) = A + Bx + Cx^2 + Dx^3 + Ex^4 + Fx^5 + Gx^6 + Hx^7,$$

we must have a

$$y' = \phi(-x) = A - Bx + Cx^2 - Dx^3 + Ex^4 - Fx^5 + Gx^6 - Hx^7.$$
$$\therefore \; y + y' = 2(A + Cx^2 + Ex^4 + Gx^6).$$

132 MENSURATION.

Therefore the whole content of the figure is

$$m \doteq \infty \sum_{n=1}^{n=m} \frac{6h}{m}\left(A + 3^2Cn^2\frac{h^2}{m^2} + 3^4En^4\frac{h^4}{m^4} + 3^6Gn^6\frac{h^6}{m^6}\right),$$

which equals

$$6h(A + 3Ch^2 + \tfrac{81}{5}Eh^4 + \tfrac{729}{7}Gh^6).$$

Now Weddle's Rule gives for the same figure

$$\begin{aligned}
5(y_2 + y_6) &= & 10(A &+ 4Ch^2 &+ 16Eh^4 &+ 64Gh^6) \\
6y_4 &= & 6A & & & \\
y_1 + y_7 &= & 2(A &+ 9Ch^2 &+ 81Eh^4 &+ 729Gh^6) \\
y_3 + y_5 &= & 2(A &+ Ch^2 &+ Eh^4 &+ Gh^6)
\end{aligned}$$

$$\therefore \xi = \tfrac{3}{10}h[20A + 60Ch^2 + 324Eh^4 + 2100Gh^6]$$

$$\therefore \xi = 6h(A + 3Ch^2 + \tfrac{81}{5}Eh^4 + \tfrac{735}{7}Gh^6).$$

Therefore the value given by our formula is in excess by the quantity $\tfrac{36}{7}Gh^7$.

Therefore the error *in the last term* is

$$\frac{\tfrac{36}{7}Gh^7}{6\times\tfrac{729}{7}Gh^7} = \tfrac{6}{729} = 0.0082+,$$

or the error is less than $\tfrac{9}{1000}$ of the last term.

Hence the error in the whole quantity will be *very much smaller* than this, since the most important terms in the expression for y are the earlier terms.

Change of origin does not change the degree of an equation; hence, we have demonstrated that by means of Weddle's Rule we can find exactly the contents of any surface or solid in which the sections may be expressed by a rational integral algebraic function, of a degree not higher than the *fifth*, of its distance from either end; and *very approximately* when the expression is of the sixth or *seventh* degree.

Cor. Suppose there are $(6n + 1)$ equidistant sections

$$y_1, \quad y_2, \quad y_3, \quad y_4, \quad \ldots, \quad y_{6n+1},$$

y_1 and y_{6n+1} being the extreme or bounding sections; and let h be their common distance; then,

$$\xi = \tfrac{3}{10}h[\Sigma y_{\text{odd}} + 5\Sigma y_{\text{even}} + \Sigma y_{\text{every third}}],$$

observing not to take either of the two extreme sections twice; that is, begin $\Sigma y_{\text{every third}}$ with y_4, and end it with y_{6n-2}.

EXAM. 98. Between y_1 and y_{19}

$$\xi = \tfrac{3}{10}h[y_1 + y_3 + y_5 + y_7 + y_9 + y_{11} + y_{13} + y_{15} + y_{17} + y_{19} +$$
$$+ 5(y_2 + y_4 + y_6 + y_8 + y_{10} + y_{12} + y_{14} + y_{16} + y_{18}) +$$
$$+ y_4 + y_7 + y_{10} + y_{13} + y_{16}].$$

EXAM. 99. Find the volume of the middle frustum of a parabolic spindle.

A *spindle* is a solid generated by the revolution of an arc of a curve round its chord, if the curve is symmetric about the perpendicular bisector of the chord. Hence a parabolic spindle is generated by the revolution of an arc of a parabola round a chord perpendicular to the axis.

Let the altitude of the middle frustum be divided into six parts each equal to h, and let A_1, A_2, A_3, A_4, A_5, A_6, A_7 be the areas of the sections.

Taking origin at center and r the longest radius, or radius of mid-section, by the equation to a parabola, we have for any point on the curve

$$x^2 = a(r - y),$$
$$\therefore y = r - \frac{x^2}{a};$$
$$\therefore \pi y^2 = \pi\left(r^2 - 2r\frac{x^2}{a} + \frac{x^4}{a^2}\right)$$

will be the area of the section at the distance x from C; and this being a rational integral function of x of the *fourth* degree, Weddle's Rule will determine the volume exactly.

Now, if r_1 is the shortest radius, or radius of the two bases, we have

$$A_1 = A_7 = \pi r_1^2;$$
$$A_2 = A_6 = \pi\left(r^2 - 2r\frac{4h^2}{a} + \frac{16h^4}{a^2}\right);$$
$$A_3 = A_5 = \pi\left(r^2 - 2r\frac{h^2}{a} + \frac{h^4}{a^2}\right);$$
$$A_4 = \pi r^2.$$

Therefore, by 101,

$$\psi = \tfrac{3}{10}h\left[5\pi\left(3r^2 - 16r\frac{h^2}{a} + 32\frac{h^4}{a^2}\right) + \pi r^2 + 2\pi r_1^2 + 2\pi\left(r^2 - 2r\frac{h^2}{a} + \frac{h^4}{a^2}\right)\right]$$

But $(3h)^2 = a(r - r_1)$; $\therefore \dfrac{h^2}{a} = \dfrac{r - r_1}{9}$.

$\therefore \psi = \tfrac{3}{10}h\pi[18r^2 + 2r_1^2 - \tfrac{28}{3}r(r - r_1) + 2(r - r_1)^2]$.

$\therefore \psi = \tfrac{3}{10}h\dfrac{\pi}{3}[32r^2 + 16rr_1 + 12r_1^2]$.

$$\therefore \psi = \tfrac{2}{5}h\pi(8r^2 + 4rr_1 + 3r_1^2). \quad Ans.$$

Making $r_1 = 0$, gives for the volume of the entire spindle,
$\tfrac{16}{5}h\pi r^2$.

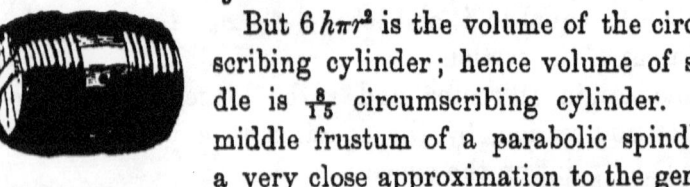
But $6h\pi r^2$ is the volume of the circumscribing cylinder; hence volume of spindle is $\tfrac{8}{15}$ circumscribing cylinder. The middle frustum of a parabolic spindle is a very close approximation to the general form of a cask, and hence is used in cask-gauging.

EXAM. 100. The interior length of a cask is 30 decimeters; the bung diameter, 24 decimeters; and the head diameters, 18 decimeters. Find the capacity of the cask.

Here $\qquad r = 12, \quad r_1 = 9, \quad h = 5;$

$\therefore k = \pi(2304 + 864 + 486) = \pi \times 3654. \quad Ans.$

EXAM. 101. The two radii which form a diameter of a circle are bisected, and ordinates are raised at the points of bisection. Find approximately the area of that portion of the circle between them.
Here
$$6h = r = y_4.$$
$$y_1 = y_7 = \sqrt{r^2 - (3h)^2} = \sqrt{r^2 - (\tfrac{1}{2}r)^2} = \sqrt{\tfrac{3}{4}r^2} = \tfrac{r}{2}\sqrt{3}.$$
$$y_2 = y_6 = \sqrt{r^2 - (2h)^2} = \sqrt{r^2 - (\tfrac{1}{3}r)^2} = \sqrt{\tfrac{8}{9}r^2} = \tfrac{r}{3}\sqrt{8}.$$
$$y_3 = y_5 = \sqrt{r^2 - h^2} = \sqrt{r^2 - (\tfrac{1}{6}r)^2} = \sqrt{\tfrac{35}{36}r^2} = \tfrac{r}{6}\sqrt{35}.$$

Hence, by 101,
$$\xi = \tfrac{3}{10} \times \tfrac{r}{6}\left[5\left(\tfrac{2r}{3}\sqrt{8} + r\right) + r + r\sqrt{3} + \tfrac{r}{3}\sqrt{35}\right].$$
$$\xi = \tfrac{r^2}{20}\left[\tfrac{10}{3}\sqrt{8} + 6 + \sqrt{3} + \tfrac{1}{3}\sqrt{35}\right] = \tfrac{r^2}{60}[18 + 20\sqrt{2} + 3\sqrt{3} + \sqrt{35}].$$

$$\therefore \xi = r^2 \times 0\cdot956608. \quad Ans.$$

But the exact area is the difference between a semicircle and the segment whose height is half the radius. Taking

$$\pi = 3\cdot1415927 \quad \text{and} \quad \sqrt{3} = 1\cdot7320508$$

gives for this area $r^2 \times 0\cdot956612$, so that our approximation is true to five places of decimals. In all approximate applications, it is desirable to avoid great differences between consecutive ordinates; applied to a quadrant of a circle, Weddle's Rule leads to a result correct to only two places of decimals.

CHAPTER VIII.

Mass-Center.

§ (X).—FOR HOMOGENEOUS BODIES.

102. The point whose distances from three planes at right angles to one another are respectively equal to the mean distances of any group of points from these planes, is at a distance from any plane whatever equal to the mean distance of the group from the same plane.

103. The mass-center of a system of equal material points is the point whose distance is equal to their average distance from any plane whatever.

104. A solid is homogeneous when any two parts of equal volume are exactly of the same mass. The determination of the mass-center of a homogeneous body is, therefore, a purely geometrical question. Again, in a very thin sheet of uniform thickness, the masses of any two portions are proportional to the areas. In a very thin wire of uniform thickness, the masses of different portions will be proportional to their lengths. Hence we may find the mass-center of a surface or of a line.

XLVII. BY SYMMETRY.

105. Two points are symmetric when at equal distances on opposite sides of a fixed point, line, or plane.

106. If a body have a plane of symmetry, the mass-center μC lies in that plane. Every particle on one side corresponds to an equal particle on the other. Hence the $^\mu C$ of every pair is in the plane, and therefore also the $^\mu C$ of the whole.

107. If a body have two planes of symmetry, the $^\mu C$ lies in their line of intersection; and if it have three planes of symmetry intersecting in two lines, the $^\mu C$ is at the point where the lines cut one another.

108. If a body have an axis of symmetry, the $^\mu C$ is in that line.

109. If a body have a center of symmetry, it is the $^\mu C$.

110. The $^\mu C$ of a straight line is its midpoint.

111. The $^\mu C$ of the circumference or area of a circle is the center.

112. The $^\mu C$ of the perimeter or area of a parallelogram is the intersection of the diagonals.

113. The $^\mu C$ of the volume or surface of a sphere is the center.

114. The $^\mu C$ of a right circular cylinder is the midpoint of its axis.

115. The $^\mu C$ of a parallelepiped is the intersection of two diagonals.

116. The $^\mu C$ of a regular figure coincides with the $^\mu C$ of its perimeter, and the $^\mu C$ of its angular points.

117. The $^\mu C$ of a trapezoid lies on the line joining the midpoints of the parallel sides.

118. Since in any triangle each medial bisects every line drawn parallel to its own base, therefore the $^\mu C$ of any triangle is the intersection of its medials. By similar triangles, this point lies on each medial two-thirds its length from its vertex, and so coincides with the $^\mu C$ of the three vertices.

119. The $^\mu C$ of the perimeter of any triangle is the center of the circle inscribed in the triangle whose vertices are the midpoints of the sides.

Proof: The mass of each side is proportional to its length, and its $^\mu C$ is its midpoint. So the $^\mu C$ of M_2 and M_3 is at a point D, such that

$$\frac{DM_2}{DM_3} = \frac{c}{b} = \frac{\tfrac{1}{2}AB}{\tfrac{1}{2}AC} = \frac{M_1 M_3}{M_1 M_2}.$$

Hence, the $^\mu C$ of the whole perimeter is in the line DM_1; and since DM_1 divides the base $M_2 M_3$ into parts proportional to the sides, it bisects $\angle M_1$. Similarly the $^\mu C$ is in the line bisecting $\angle M_2$.

120. The $^\mu C$ of the surface of any tetrahedron is the center of the sphere inscribed in the tetrahedron whose summits are the $^\mu C$'s of the faces.

121. If the vertex of a triangular pyramid be joined with the $^\mu C$ of the base, the $^\mu C$ of the pyramid is in this line at three-fourths of its length from the vertex. (Proof by similar triangles.)

122. The $^\mu C$ of a tetrahedron is also the $^\mu C$ of its four summits.

123. The $^\mu C$ of any pyramid or cone is in the line joining the $^\mu C$ of the base with the apex at three-fourths of its length from the apex.

XLVIII. The Mass-Center of a Quadrilateral.

124. A *sect* is a limited line or rod.

125. The *opposite* to a point P on a sect AB is a point P', such that P and P' are at equal distances from the center of AB, but on opposite sides of it.

126. The $^\mu C$ of any quadrilateral is the $^\mu C$ of the triangle whose apices are the intersection of its two diagonals and the *opposites* of that intersection on those two diagonals respectively.

Proof: Construct the $^\mu C$'s E and F of the triangles ABC and ACD made by the diagonal AC of the quadrilateral $ABCD$; then the point on the line EF, which divides it inversely as the areas of these triangles, will be the $^\mu C$ of the quadrilateral. But if the diagonal BD is cut by AC in the point G, then $ABC:ACD = BG:GD$. So the sought point is the $^\mu C$ of $BG \times E$ and $DG \times F$; that is, of $BG \times A$, $BG \times B$, $BG \times C$, and $GD \times A$, $GD \times D$, $GD \times C$. But we may substitute $BD \times A$ for $BG \times A$ and $GD \times A$; also $BD \times C$ for $BG \times C$ and $GD \times C$. For $BG \times B$ and $GD \times D$ we may substitute $BD \times K$, where K is the *opposite* of G on the sect BD. Therefore the sought point is the $^\mu C$ of A, C, and K, that is, of G, J, and K, where J is the opposite of G on AC.

Cor. Calling the $^\mu C$ of the quadrilateral L, we have

$$ACL = ACF - AEC = \tfrac{1}{3}(ACD - ABC).$$

127. The $^\mu C$ of the four angular points of a quadrilateral is the intersection of the lines joining the midpoints of pairs of opposite sides. Let this point O be called the midcenter, and G the intersection of the two diagonals be

140   MENSURATION.

called the cross-center; then the $^\mu C$ of the quadrilateral is in the line joining these two centers produced past the mid-center, and at a distance from it equal to one-third of the distance between the two centers.

That is, LOG will be a straight line, and

$$OL = \tfrac{1}{3} OG.$$

128.   THE MASS-CENTER OF AN OCTAHEDRON.

Let AF, BG, CH be three sects (finite lines) not meeting. By an *octahedron* understand the solid whose eight faces are ABC, ACG, AGH, AHB, FBC, FCG, FGH, FHB. The solid is girdled by the perimeters of three skew quadrilaterals, $BCGH$, $CAHF$, $ABFG$. Now the mid-points of the sides of any skew quadrilaterals are in one plane. Draw, then, three planes bisecting the sides of these quadrilaterals, and let them meet in a point K, which call the *cross-center*.

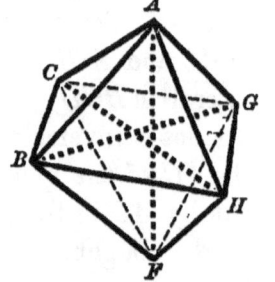

Let, also, M (*mid-center*) be the mean point of the six vertices A, B, C, F, G, H; it is the $^\mu C$ of the triangle formed by the mid-points of AF, BG, CH. To find S, the $^\mu C$ of the solid, join KM, and produce it to S so that $MS = \tfrac{1}{3} KM$.

Proof: The solid is the sum of the four tetrahedra $AFBC$, $AFCG$, $AFGH$, $AFHB$.

Now the $^\mu C$ of a tetrahedron is the $^\mu C$ or mean point of its vertices; consequently the line joining the $^\mu C$ of $AFBC$ to the mid-point of GH is divided by the point M in the ratio $1:2$. The same is true of the other three tetrahedra and the mid-points of HB, BC, CG. Hence, the mass-centers of the four tetrahedra are in one plane passing through the point S found by the above construction, and therefore the $^\mu C$ of the whole solid is in this plane. So, also, it is in the other two planes determined by dividing the solid into tetrahedra having the common edge BG and the common edge CH respectively. Therefore it coincides with the point S.

Cor. By making the pairs of faces ABH, AHG; ACG, CFG; CBF, BHF to be respectively coplanar, we pass to a truncated triangular pyramid. If we join its cross-center K with its mïd-center M, and produce KM to S, making $MS = \frac{1}{2} KM$, then S will be the $^\mu C$ of the trunc.

NOTE. This corollary was Sylvester's extension of the geometric method of centering the plane quadrilateral, and suggested to Clifford the above.

XLIX. GENERAL MASS-CENTER FORMULA.

In any body between parallel planes, we can reckon the distance of its $^\mu C$ above its base, if every cross-section is a given function $\phi(x)$ of its distance x from the base.

The prism whose base is the section $\phi(x)$, and whose height is the nth part of the altitude a of the body, has for volume $\frac{a}{n}\phi(x)$. Its $^\mu C$ is $x+\frac{a}{2n}$ from the base of the body, and has the coefficient $\frac{a}{n}\phi(x)$. Now suppose the $^\mu C$ of the body distant τ from its base, and form the product of τ with the sum D of the values which $\frac{a}{n}\phi(x)$ takes when x equals $0, \frac{1}{n}a, \frac{2}{n}a, \ldots, \frac{n-1}{n}a$; also form the sum S of the values which $\left(x+\frac{a}{2n}\right)\frac{a}{n}\phi(x)$ takes for the same worths of x. The product τD equals the sum S when the arbitrary number n is taken indefinitely great.

$$\tau D = S, \quad \text{for} \quad n \doteq \infty.$$

But for $n \doteq \infty$, the sum D expresses the volume of the body.

The sum S consists of the sum C of terms from $\dfrac{a}{n}x\phi(x)$, and of the sum E of terms from $\dfrac{a}{2n}\cdot\dfrac{a}{n}\phi(x)$. But the sum E has the value $\dfrac{a}{2n}D$, and vanishes for $n \doteq \infty$.

Therefore, for the determination of τ, we have the equation

$$\tau D = C.$$

129.   Mass-Center of any Prismatoid.

If $\phi(x)$ is of degree not higher than the second, then $x\phi(x)$, which we will call $f(x)$, is not higher than the third. Therefore, by 99,

$$D = \tfrac{1}{6}a[\phi(0) + 4\phi(\tfrac{1}{2}a) + \phi(a)].$$
$$C = \tfrac{1}{6}a[f(0) + 4f(\tfrac{1}{2}a) + f(a)].$$

But $\quad f(0) = 0, \quad f(\tfrac{1}{2}a) = \tfrac{1}{2}a\phi(\tfrac{1}{2}a), \quad f(a) = a\phi(a).$

Therefore, $\quad \dfrac{\tau}{a} = \dfrac{2\phi(\tfrac{1}{2}a) + \phi(a)}{\phi(0) + 4\phi(\tfrac{1}{2}a) + \phi(a)}.$

Therefore, $\quad \tau = \tfrac{1}{2}a + \dfrac{a^2[\phi(a) - \phi(0)]}{12D},$

or $\quad \tau = \tfrac{1}{2}a + \dfrac{a^2(B_2 - B_1)}{12D}.$

130. For the applicability of the Mass-Center Formula:

$$\tau = \tfrac{1}{2}a + \dfrac{a^2(B_2 - B_1)}{12D},$$

the *test* is

$$A_x = q + mx + nx^2.$$

For an examination of the possible varieties, see 99, § (T), (1), (2), (3), (5), (6), (8), (9); and § (U), (1), (2), (3), (5), (6), (7), (8).

Of course it applies also to the corresponding plane figures.

Exam. 102. For trapezoid

$$\tau = \tfrac{1}{2}a + \frac{a^2(b_2 - b_1)}{12(b_1 + b_2)\tfrac{1}{2}a} = \frac{a(2b_2 + b_1)}{3(b_1 + b_2)}.$$

Exam. 103. For pyramid or cone

$$\tau = \tfrac{1}{2}a + \frac{a^2 B_2}{12(\tfrac{1}{3}aB_2)} = \tfrac{3}{4}a$$

from B_1, or $\tfrac{1}{4}a$ from B_2.

Exam. 104. The $^\mu C$ of a pyramidal frustum is in the line joining the $^\mu C$'s of the parallel faces, and

$$\tau = \tfrac{1}{2}a + \frac{a(B_2 - B_1)}{4(B_1 + \sqrt{B_1 B_2} + B_2)}.$$

So, if any two homologous basal edges are as l to λ, the distance of the frustral $^\mu C$ from one base will be $\dfrac{a}{4}\left(\dfrac{l^2 + 2l\lambda + 3\lambda^2}{l^2 + l\lambda + \lambda^2}\right)$, and from the other $\dfrac{a}{4}\left(\dfrac{\lambda^2 + 2l\lambda + 3l^2}{l^2 + l\lambda + \lambda^2}\right)$.

Exam. 105. From Exam. 95, for paraboloid

$$\tau = \tfrac{1}{2}a + \frac{a^2 \pi r^2}{6\pi r^2 a} = \tfrac{2}{3}a.$$

For frustum of this we obtain an expression similar to that for trapezoid. It is

$$\tau = \tfrac{1}{2}a + \frac{a^2 \pi (r_2^2 - r_1^2)}{\tfrac{1}{2}a \cdot 2\pi(r_2^2 + r_1^2)} = \frac{a(2r_2^2 + r_1^2)}{3(r_2^2 + r_1^2)}.$$

Exam. 106. For $^\mu C$ of half-globe, from center,

$$\tau = \tfrac{1}{2}r - \frac{\pi r^4}{12(\tfrac{2}{3}\pi r^3)} = \tfrac{3}{8}r.$$

131. The average haul of a piece of excavation is the distance between the $^\mu C$ of the material as found and its $^\mu C$ as deposited.

132. The $^\mu C$ of a series of consecutive equally-long quadratic shapes may be found by assuming the $^\mu C$ of each shape to be in its mid-section, then compounding, and to the distance of the point thus found from B_1 adding $\dfrac{a^2(B_2-B_1)}{12D}$.

Note. It is a singular advantage of the $^\mu C$ formula that its second or correction term remains as simple for any number of shapes in the series as for one.

In consequence, the error of the assumption that the $^\mu C$ of each shape is in its mid-section, is less as the series is longer. No error whatever results from this assumption when the end areas B_1 and B_2 are equal. For instance, in finding $^\mu C$ of a spherical sector whose component cone and segment have equal altitude, we may assume that the $^\mu C$ of each is midway between its bases.

EXERCISES AND PROBLEMS IN MENSURATION.

Problems and Exercises on Chapter I.

1. $a^2 + b^2 = c^2.$

EXERCISE 1. Find the diagonal of a square whose side is unity. $\sqrt{2} = 1{\cdot}41421+.$ *Answer.*

2. Find the diagonal of a cube whose edge is unity. $\sqrt{3} = 1{\cdot}732050+.$ *Ans.*

3. To draw a perpendicular to a line at the point C.

Measure $CA = 3$ meters, and fix at A and C the ends of a line 9 meters long; which stretch by the point B taken 4 meters from C and 5 meters from A.

BC is $\perp$ to AC because $3^2 + 4^2 = 5^2$.

4. The whole numbers which express the lengths of the sides of a right-angled triangle, when reduced to the lowest numbers possible by dividing them by their common divisors, cannot be all even numbers. *Nor can they be all odd.*

For, if a and b are odd, a^2 and b^2 are also odd, each being an odd number taken an odd number of times.

$a^2 + b^2$ is even, and $\therefore$ c is even.

2. $c^2 - a^2 = (c + a)(c - a) = b^2.$

5. To obtain three whole numbers which shall represent the sides of a right-angled triangle.

Rule of Pythagoras. Take n any odd number, then $\dfrac{n^2-1}{2}$ = the second number, and $\dfrac{n^2+1}{2}$ = the third number.

Plato's Rule. Take any even number m, then also $\dfrac{m^2}{4}-1$, and $\dfrac{m^2}{4}+1$.

Euclid's Rule. Take x and y, either both odd or both even, such that $xy = b^2$, a perfect square; then $a = \dfrac{x-y}{2}$ and $c = \dfrac{x+y}{2}$.

Rule of Maseres. Of any two numbers take twice their product, the difference of their squares, and the sum of their squares.

Proof: Let the whole numbers which express the lengths of the two sides of a rt. $\triangle$ be a and m. If c be a whole number, it must $= a + n$, where n is a whole number.

By 1,
$$\therefore a^2 + m^2 = c^2 = a^2 + 2an + n^2.$$
$$\therefore m^2 = 2an + n^2.$$
$$\therefore 2an = m^2 - n^2.$$
$$\therefore a = \frac{m^2 - n^2}{2n}.$$
$$\therefore a + n = c = \frac{m^2 - n^2 + 2n^2}{2n} = \frac{m^2 + n^2}{2n}.$$

Therefore, the three sides, a, m, c, are $\dfrac{m^2-n^2}{2n}$, $\dfrac{2mn}{2n}$, and $\dfrac{m^2+n^2}{2n}$. Magnifying the rt. $\triangle$ $2n$ times, its sides are expressed by the whole numbers $m^2 - n^2$, $2mn$, and $m^2 + n^2$.

EXERCISES AND PROBLEMS. 147

Table I.—Dissimilar Right-Angled Triangles.

Sides.			Area.	Sides.			Area.
3	4	5	6	57	176	185	5,016
5	12	13	30	85	132	157	5,610
8	15	17	60	36	323	325	5,814
7	24	25	84	29	420	421	6,090
9	40	41	180	60	221	229	6,630
12	35	37	210	119	120	169	7,140
20	21	29	210	31	480	481	7,440
11	60	61	330	84	187	205	7,854
16	63	65	504	104	153	185	7,956
13	84	85	546	95	168	193	7,980
28	45	53	630	40	399	401	7,980
15	112	113	840	69	260	269	8,970
33	56	65	924	33	544	545	8,976
20	99	101	990	68	285	293	9,690
17	144	145	1,224	133	156	205	10,374
48	55	73	1,320	44	483	485	10,626
36	77	85	1,386	35	612	613	10,710
39	80	89	1,560	105	208	233	10,920
19	180	181	1,710	75	308	317	11,550
24	143	145	1,716	96	247	265	11,856
21	220	221	2,310	140	171	221	11,970
65	72	97	2,340	120	209	241	12,540
44	117	125	2,574	37	684	685	12,654
60	91	109	2,730	76	357	365	13,566
28	195	197	2,730	48	575	577	13,800
23	264	265	3,036	115	252	277	14,490
51	140	149	3,570	39	760	761	14,820
25	312	313	3,900	52	675	677	17,550
32	255	257	4,080	87	416	425	18,096
52	165	173	4,290	160	231	281	18,480
88	105	137	4,620	136	273	305	18,564
27	364	365	4,914	161	240	289	19,320

Table I.—Continued.

Sides.			Area.	Sides.			Area.
56	783	785	21,924	319	360	481	57,420
93	476	485	22,134	124	957	965	59,334
207	224	305	23,184	231	520	569	60,060
120	391	409	23,460	200	609	641	60,900
135	352	377	23,760	279	440	521	61,380
92	525	533	24,150	185	672	697	62,160
175	288	337	25,200	336	377	505	63,336
204	253	325	25,806	80	1,599	1,601	63,960
152	345	377	26,220	308	435	533	66,990
180	299	349	26,910	195	748	773	72,930
60	899	910	26,970	396	403	565	79,794
145	408	433	29,580	259	660	709	85,470
225	272	353	30,600	368	465	593	85,560
100	621	629	31,050	336	527	625	88,536
105	608	617	31,920	315	572	653	90,090
189	340	389	32,130	273	736	785	100,464
64	1,023	1,025	32,736	400	561	689	112,200
252	275	373	34,650	364	627	725	114,114
168	425	457	35,700	455	508	697	120,120
155	468	493	36,270	407	624	745	126,984
228	325	397	37,050	301	900	949	135,450
111	680	689	37,740	468	595	757	139,230
108	725	733	39,150	432	665	793	143,640
68	1,155	1,157	39,270	369	800	881	147,600
203	396	445	40,194	429	700	821	150,150
165	532	557	43,890	315	988	1,037	155,610
297	304	425	45,144	555	572	797	158,730
72	1,295	1,297	46,620	540	629	829	169,830
184	513	545	47,196	451	780	901	175,890
116	837	845	48,546	504	703	865	177,156
280	351	449	49,140	329	1,080	1,129	177,660
217	456	505	49,476	420	851	949	178,710
261	380	461	49,590	464	777	905	180,264
76	1,443	1,445	54,834	533	756	925	201,474

Table I.—Concluded.

Sides.			Area.	Sides.			Area.
616	663	905	204,204	748	1,035	1,277	387,090
473	864	985	204,336	893	924	1,285	412,566
580	741	941	214,890	560	1,551	1,649	434,280
496	897	1,025	222,456	884	987	1,325	436,254
615	728	953	223,860	792	1,175	1,417	465,300
330	644	725	107,226	684	1,363	1,525	466,830
557	840	1,009	234,780	740	1,269	1,469	469,530
696	697	985	242,556	931	1,021	1,381	474,810
660	779	1,021	257,070	833	1,056	1,345	481,474
645	812	1,037	261,870	969	1,120	1,481	542,640
476	1,107	1,205	263,466	720	1,519	1,681	546,840
620	861	1,061	266,910	836	1,325	1,565	554,014
585	928	1,097	271,440	780	1,421	1,621	554,190
731	780	1,069	285,090	936	1,127	1,465	556,738
504	1,247	1,345	314,244	1,036	1,173	1,565	607,614
704	903	1,145	317,856	988	1,275	1,613	629,850
660	989	1,189	326,370	880	1,479	1,721	650,760
612	1,075	2,237	329,950	1,113	1,184	1,625	658,896
765	868	1,157	332,010	1,140	1,219	1,669	694,830
705	992	1,217	349,680	1,040	1,431	1,769	744,120
832	855	1,193	355,680	1,248	1,265	1,777	799,480
532	1,395	1,493	371,070	1,148	1,485	1,877	852,390
799	960	1,249	383,520	1,312	1,425	1,937	863,550

Sides.			Area.
4,059	4,060	5,741	
23,660	23,661	33,461	
207,000	207,151	292,849	
159,140,519	159,140,520	225,058,681	

3. $a^2 + b^2 + 2bj = c^2$.

6. Two sides, $a = 6{\cdot}708$, $b = 5$, contain an obtuse angle. If $j = 3$, find c.   10. *Ans.*

7. If $a = 13$, $b = 11$, $c = 20$, find j.   5. *Ans.*

4. $a^2 + b^2 - 2bj = c^2$.

8. The three sides of a triangle are 2·5 meters, 4·8 meters, 3·2 meters; find the projections of the other two sides on $b = 4{\cdot}8$ meters.

     1·985 meters and 2·815 meters. *Ans.*

9. Two sides of a triangle, 3 meters and 8 meters long, enclose an angle of 60°. Find the third side.

     7 meters. *Ans.*

HINT. Joining the midpoint of 3 with the vertex of the rt. $\angle$ made in projecting 3 on 8 gives two isosceles triangles, and $\therefore j = 1\tfrac{1}{2}$.

10. Two sides of a triangle are 13 meters and 15 meters. Opposite the first is an angle of 60°. Find the third.

     7 or 8 meters. *Ans.*

HINT. Drop $\perp$ to 15-meter side. The segment adjacent the third side is half the third side.

11. Two sides of a triangle are 9·6 meters and 12·8 meters; the perpendicular from their vertex on the third side is 7·68 meters; find that side.   4. *Ans.*

6. $a^2 + c^2 - \tfrac{1}{2} b^2 = 2i^2$.

12. Given i_1, the medial from A to a; i_2, from B to b; and i_3, from C to c. Find a, b, and c.

From 6,
$$a^2 + c^2 - \tfrac{1}{2} b^2 = 2i_2^2,$$
$$a^2 + b^2 - \tfrac{1}{2} c^2 = 2i_3^2,$$
$$b^2 + c^2 - \tfrac{1}{2} a^2 = 2i_1^2.$$

Taking the last equation from twice the sum of the two former gives
$$4a^2 + \tfrac{1}{2}a^2 = 2(2i_2^2 + 2i_3^2 - i_1^2).$$
$$\therefore a = \tfrac{2}{3}\sqrt{2i_2^2 + 2i_3^2 - i_1^2}. \ Ans.$$

Symmetrically, find b and c.

13. If $i_1 = 18$, $i_2 = 24$, $i_3 = 30$; find a, b, and c.
$$a = 34 \cdot 176, \ b = 28 \cdot 844, \ c = 20. \ Ans.$$

14. Prove $i_1^2 + i_2^2 + i_3^2 = \tfrac{3}{4}(a^2 + b^2 + c^2)$.

15. In any right-angled triangle prove $\tfrac{1}{2}a = \sqrt{\dfrac{4i_2^2 - i_1^2}{15}}$.

16. The locus of a point, the sum of the squares of whose distances from two fixed points is constant, is a circumference whose center is the midpoint of the straight line joining the two fixed points.

17. The locus of points, the difference of the squares of whose distances from two fixed points is constant, is a straight line perpendicular to that which joins the two fixed points.

18. The sum of any two sides of a triangle is greater than twice the concurrent medial.

19. In every quadrilateral the sum of the squares of the four sides exceeds the sum of the squares of the two diagonals by four times the square of the straight line joining the middle points of the diagonals.

20. The sum of the squares on the four sides of a parallelogram is equivalent to the sum of the squares on the diagonals.

21. The sum of the squares of the diagonals of a trapezoid is equal to the sum of the squares of the non-parallel sides augmented by twice the product of the parallel sides.

22. On the three sides of a triangle squares are described outward. Prove that the three lines joining the ends of their outer sides are twice the medials of the triangle and perpendicular to them.

23. Prove that the medials of a triangle cut each other into parts which are as 1 to 2.

24. The intersection-point of medials is the center of mass of the triangle.

25. Prove an ∡ in a △, acute, rt., or obtuse, according as the medial through the vertex of that ∡ is $>$, $=$, or $<$ half the opposite side.

7. $b_2 = \dfrac{a_2 b_1}{a_1}$.

26. The three sides of a triangle are 17·4 meters, 23·4 meters, 31·8 meters. The smallest side of a similar triangle is 5·8 meters. Find the other sides.
7·8 meters and 10·6 meters. *Ans.*

27. Find the height of an object whose shadow is 37·8 meters, when a rod of 2·75 meters casts 1·4 meters shadow.
74·25 meters. *Ans.*

28. The perpendicular from any point on a circumference to the diameter is a mean proportional between the two segments of the diameter.

29. Every chord of a circle is a mean proportional between the diameter drawn from one of its extremities, and its projection on that diameter.

30. From a hypothenuse of 72·9 meters, a perpendicular from the right angle cuts a part equal to 6·4 meters; find the sides. $a = 21·6$ meters, $b = 69·62$ meters. *Ans.*

EXERCISES AND PROBLEMS. 153

31. The hypothenuse is 32·5 meters, and the perpendicular on it 15·6 meters; find the segments.
11·7 and 20·8. *Ans.*

32. In a rt. $\triangle$, $c = 36\cdot5$ meters; $a + b = 51\cdot1$ meters; find a and b. $a = 21\cdot9$ meters, $b = 29\cdot2$ meters. *Ans.*

33. The sides of a triangle are 4·55, 6·3, and 4·445; the perimeter of a similar triangle is 4·37. Find its sides.
1·3, 1·8, and 1·27. *Ans.*

34. A lamp-post 3 meters high is 5 meters from a man 2 meters tall; find the length of his shadow.

35. The sides of a pentagon are 12, 20, 11, 15, and 22; the perimeter of a similar pentagon is 16 meters. Find its sides.   2·4, 4, 2·2, 3, and 4·4. *Ans.*

36. Through the point of intersection of the diagonals of a trapezoid a line is drawn parallel to the parallel sides. Prove that the parallel sides have the same ratio as the parts into which this line cuts the non-parallel sides.

8. $d = \dfrac{k_{\frac{1}{2}}^2}{h}$.

9. $h = r - \sqrt{r^2 - \frac{1}{4}k^2}$.

10. $k_{\frac{1}{2}} = \sqrt{h^2 + \frac{1}{4}k^2}$.

37. Find the side of a regular decagon.

If a radius is divided in extreme and mean ratio, the greater segment is equal to a side of the inscribed regular decagon.   Ww. 394; (Eu. IV. 10; Cv. V. 17).

$$\therefore r(r - k_{10}) = k_{10}^2,$$
$$r^2 - rk = k^2,$$
$$k^2 + rk = r^2,$$
$$k^2 + rk + \tfrac{1}{4}r^2 = \tfrac{5}{4}r^2.$$

$$\therefore k_{10} = r\frac{\sqrt{5} - 1}{2}. \ \textit{Ans.}$$

11. $k_4 = \sqrt{2r^2 - r\sqrt{4r^2 - k^2}}$.

38. Prove that the sides of the regular pentagon, hexagon, and decagon will form a right-angled triangle; or $k_{10}^2 + k_6^2 = k_5^2$.

39. Show that $k_{24} = r\sqrt{2 - \sqrt{2 + \sqrt{3}}}$.

$$k_{48} = r\sqrt{2 - \sqrt{2 + \sqrt{2 + \sqrt{3}}}},$$
$$k_{96} = r\sqrt{2 - \sqrt{2 + \sqrt{2 + \sqrt{2 + \sqrt{3}}}}}, \text{ etc.}$$

40. Show that $k_4 = r\sqrt{2}$.

$$k_8 = r\sqrt{2 - \sqrt{2}},$$
$$k_{16} = r\sqrt{2 - \sqrt{2 + \sqrt{2}}},$$
$$k_{32} = r\sqrt{2 - \sqrt{2 + \sqrt{2 + \sqrt{2}}}}, \text{ etc.}$$

12. $t = \dfrac{2kr}{\sqrt{4r^2 - k^2}}$.

41. If a_n is the apothegm of a regular inscribed n-gon, prove $a_{2n} = \sqrt{\dfrac{r(r + a_n)}{2}}$.

HINT. Use 7 and 2.

42. If also p'_n be perimeter of inscribed polygon, p_n of circumscribed, prove $p'_{2n} = \dfrac{p'_n r}{a_{2n}}$.

43. $p_n : p'_n :: r : a_n$.

44. Indicate how to reckon π by 40, 41, and 42, or by the following $p_{2n} = \dfrac{2 p'_n p_n}{p'_n + p_n}$ and $p'_{2n} = \sqrt{p'_n p_{2n}}$.

13.

45. When a quantity increases continually without becoming greater than a fixed quantity, it has a limit equal or inferior to this constant.

EXERCISES AND PROBLEMS. 155

46. When a quantity decreases continually without becoming less than a fixed quantity, it has a limit equal or superior to this constant.

47. When two variables are always equal, if one of them has a limit, so has the other also.

14. Lim $p_n = \lim p'_n = c$.

48. Show that $p'_{2n}{}^2 = p'_n p_{2n}$.

49. Prove $p'_{2n} = p'_n \sqrt{\dfrac{2p_n}{p_n + p'_n}}$, and also $p_{2n} = \dfrac{2 p_n p'_n}{p_n + p'_n}$.

50. If $\Re_n$, r_n be the radii of the circumscribed and inscribed circles of a regular polygon of n sides, and $\Re_{2n}$, r_{2n} the corresponding radii for a regular $2n$-gon of the *same perimeter* as the n-gon; then $\Re_n r_{2n} = \Re_{2n}{}^2$, and $\Re_n + r_n = 2 r_{2n}$.

51. If $\measuredangle_1, \measuredangle_2, \measuredangle_3$, etc., $\measuredangle_{2n}$ be the angles of a polygon of $2n$ sides, inscribed in a circle, then $\measuredangle_1 + \measuredangle_3 +$ etc. $+ \measuredangle_{2n-1}$ $= \measuredangle_2 + \measuredangle_4 +$ etc. $+ \measuredangle_{2n}$.

52. The greater the number of sides of a regular polygon, the greater is the magnitude of each of its angles. The limit is f, which can never be reached since the sum of the exterior angles is always 6.

15. $\pi = 3{\cdot}14159265358979323846264338\,3279+$.

16. $c_1 : c_2 :: r_1 : r_2$.

17. $\pi = \dfrac{c}{d} = \dfrac{\frac{1}{2}c}{r} = 3{\cdot}1416-$.

18. $c = d\pi = 2r\pi$.

53. Find c when $r = 14$, taking $\pi = \frac{22}{7}$.   88. *Ans*.

19. $d = 2r = \dfrac{c}{\pi} = c \times 0{\cdot}3183098+$.

54. Find the diameter of a wheel, which, in a street 19,635 meters long, makes 3,125 revolutions.   2 meters. *Ans*.

156 MENSURATION.

55. The hypothenuse is 10, and one side is 8; semicircles are described on the three sides. Find the radius of the semicircle whose circumference equals the circumferences of the three semicircles so described. 12. *Ans.*

56. Find the radius of a sphere in which a section 30 centimeters from the center has a circumference of 251·2 centimeters.
$$r = \sqrt{3^2 + \left(\frac{251\cdot 2}{2 \times \pi}\right)^2}. \text{ Ans.}$$

Exercises and Problems on Chapter II.

20. *Arc measures ⦨ at center.*

57. Find the third angle of a triangle whose first angle is 12° 56″ and second angle 114° 48″. 53° 58′ 16″. *Ans.*

58. The angles of a triangle are as $1:2:3$. Find each.
30°, 60°, 90°. *Ans.*

59. Of the three angles of a triangle ⦨ a is 12° 20′ smaller than ⦨ β, and ⦨ γ is 5° 43′ smaller than ⦨ β. Find each. $a = 53° 41′$, $\beta = 66° 1′$, $\gamma = 60° 18′$. *Ans.*

60. The sum of two angles of a triangle is 174° 48′ 24″; the difference of the same two is 48° 24′ 50″. Find all three. 110° 36′ 37″, 63° 11′ 47″, 5° 11′ 36″. *Ans.*

61. Three angles of a quadrilateral are 125° 48′ 32″, 127° 58′ 45″, 85° 37′ 27″; find the fourth. 20° 35′ 16″. *Ans.*

62. The angles of a quadrilateral are as $2:3:4:7$; find each. 45°, 67° 30′, 90°, 157° 30′. *Ans.*

63. In what regular polygon is every angle 168° 45′?
A 32-gon. *Ans.*

64. The vertical angle of an isosceles triangle is 148° 47″; find the others. 15° 59′ 36½″. *Ans.*

EXERCISES AND PROBLEMS. 157

65. The sum of two angles adjacent to one side of an isosceles triangle is 35°23′48″; find the three angles.
44°36′12″, 44°36′12″, and 90°47′36″. *Ans.*

21. $u = \dfrac{l}{r}$.

66. Find the circular measure of the angle subtended by a circular watch-spring 3 millimeters long and radius 1½ millimeters.

67. If a perigon be divided into n equal parts, how many of them would a radian contain? $\dfrac{n}{2\pi}$. *Ans.*

22. $l = \dfrac{cg°}{360°}$.

68. Find the arc pertaining to a central angle of 78° when $r = 1.5$ meters. $l = 2.042$ meters. *Ans.*

69. Find the arc intercepted by a central angle of 36°25′ when $r = 8.5$ meters. $l = 5.39$ meters. *Ans.*

70. Find an arc of 112° which is 4 meters longer than its radius. $l = 8.189$ meters. *Ans.*

23. $g° = \dfrac{l360°}{c}$.

71. When $d = 11.5$, find arc 4.6 meters long.
∡ $= 45°57′14¼″$. *Ans.*

72. Calling $\pi = \tfrac{22}{7}$, find r when 64° measure 70.4 meters.
$r = 63$ meters. *Ans.*
Hint. $2\pi r : 70.4 :: 360 : 64$.

24, 25, 26, 27, and 28.

73. Find the complement, supplement, and explement of 30°.

MENSURATION.

74. Find the angle between the bisectors of any two adjacent angles.

75. If a medial equals half its base, its angle is right.

76. If one angle in a right-angled triangle is 30°, one side is half the hypothenuse.

77. In every isosceles right triangle half the hypothenuse equals the altitude upon it.

78. One angle in a right triangle is 30°; into what parts does the altitude divide the right angle?

79. How large is the angle between perpendiculars on two sides of an equilateral triangle?

80. Find the inscribed angle standing on an arc of 116°27′38″. 58°13′49″. *Ans.*

81. Find the inscribed angle cutting out one-tenth circumference. 18°. *Ans.*

82. Find the angle of intersection of two secants which include arcs of 100°48′ and 54°12′. 23°18′. *Ans.*

83. An angle made by two tangents is measured by the difference between 180° and the smaller intercepted arc.

84. From the same point in a circumference two chords are drawn which cut off respectively arcs of 120° and 80°; find the included angle.

85. The four angle-bisectors of any quadrilateral from a quadrilateral whose opposite angles are supplemental.

29. $u = \dfrac{\pi g°}{180°}$.

86. Find the circular measure of 42°. ·73303. *Ans.*

87. Find the circular measure of 45°. ·785398. *Ans.*

EXERCISES AND PROBLEMS. 159

88. Find the circular measure of 30°. ·523598. *Ans.*

89. Find the circular measure of 60°. $\dfrac{\pi}{3}$. *Ans.*

90. Find the circular measure of $\pi°$. $\dfrac{\pi^2}{180}$. *Ans.*

91. Express seven-sixteenths of a right angle in circular measure.

92. Express in circular measure an angle of 240°.
$\dfrac{4\pi}{3}$. *Ans.*

93. Find in circular measure the angle made by the hands of a watch at 5:15 o'clock.

94. Find u of ∡ made by watch-hands at a quarter to 8.

95. Find u of watch ∡ at 3:30 o'clock.

96. Find u of watch ∡ at 6:05 o'clock.

97. The length of an arc of 45° in one circle is equal to that of 60° in another. Find the circular measure of an angle which would be subtended at the center of the first by an arc equal to the radius of the second.   ¾. *Ans.*

30. $g° = \dfrac{u\,180°}{\pi}$.

98. The angle whose circular measure equals one-half is 28° 38′ 52″ 24‴.

99. Find the number of degrees in an angle whose $u = ⅔$.
$g° = ⅔(57·2957795+) = 38·197186+$. *Ans.*

100. Find ∡ whose $u = ¾$.   42°·9718346. *Ans.*

101. Find the number of degrees in an angle whose $u = ⁵⁄₄$.   71·61972439. *Ans.*

102. Find the number of degrees, minutes, and seconds in an angle whose $u = \frac{11}{21}$. $30° 0' 43''\cdot 45$. *Ans.*

103. The circular measure of the sum of two angles is $\frac{5}{12}\pi$, and their difference is $17°$; find the angles.
$$27° 15' \text{ and } 10° 15'. \textit{ Ans.}$$

104. Express in degrees the angle whose $u = \frac{2}{3}\pi$.
$$120°. \textit{ Ans.}$$

105. How many degrees, minutes, and seconds are there in the angle whose circular measure is $\frac{5}{6}$?
$$47° 44' 47'' \text{ nearly. } \textit{Ans.}$$

106. Express in degrees and circular measure the vertical angle of an isosceles triangle which is half of each of the angles at the base.

107. How many times is the angle between two consecutive sides of a regular hexagon contained (1) in a right angle? (2) in a radian?   (1) $\frac{3}{4}$, (2) $\frac{3}{2\pi}$. *Ans.*

108. Two wheels with fixed centers roll upon each other, and the circular measure of the angle through which one turns gives the number of degrees through which the other turns in the same time; find the ratio of the radii of the wheels. $\frac{180}{\pi}$. *Ans.*

31. $r = \dfrac{l 180°}{\pi g°}$.

109. The length of an arc of $60°$ is $36\frac{2}{3}$; find the radius.
$$r = 35. \textit{ Ans.}$$

110. Find circumference where $\angle\, 30°$ is subtended by arc 4 meters.

111. If l_1 be the length of an arc of $45°$ to radius r_1, and l_2 the length of arc $60°$ to r_2, prove $3 r_1 l_2 = 4 r_2 l_1$.

EXERCISES AND PROBLEMS ON CHAPTER III.

32. $R = ab$.

112. The area of a rectangle is 2,883 square meters; the diagonal measures 77·5 meters. Find the sides.
$$b = 46\cdot5 \text{ meters}, \quad a = 62 \text{ meters. } Ans.$$

HINT.
$$a^2 + b^2 = (77\cdot5)^2 \quad \ldots \ldots \quad \text{I.}$$
$$ab = 2{,}883 \quad \ldots \ldots \quad \text{II.}$$

Substitute in I. the value of a from II. This gives a biquadratic soluble as a quadratic. Or, to avoid the solution of a quadratic, multiply II. by 2, then add and subtract from I. This gives

$$a^2 + 2ab + b^2 = 11{,}772\cdot25,$$
$$a^2 - 2ab + b^2 = \phantom{11{,}7}240\cdot25.$$
$$\therefore a + b = \phantom{11{,}7}108\cdot5,$$
$$a - b = \phantom{11{,}77}15\cdot5.$$

113. Find the perimeter and area of a rectangle whose altitude $a = 1{,}843\cdot02$ meters and base $b = 845\cdot6$ meters.
$$p = 5{,}377\cdot24 \text{ meters},$$
$$R = 1{,}558{,}457\cdot712 \text{ square meters. } Ans.$$

114. Find the number of boards 4 meters long, 0·5 meters broad, necessary to floor a rectangular room 16 meters long and 8 meters broad. 64. *Ans.*

115. Of two equivalent rectangles, one is 4·87 meters long and 2·84 meters broad, the other is 4·26 meters long. How broad is it? 3·246 meters. *Ans.*

116. The perimeter of a rectangle is 24·54 meters; the base is double the altitude. Find the area.
$$R = 33\cdot4562 \text{ square meters. } Ans.$$

117. The difference of two sides of a rectangle is 1·4 meters; their sum, 8·2 meters. Find its area.
$$R = 16\cdot32 \text{ square meters. } Ans.$$

118. The base of a rectangle of 46·44 square meters is 3·2 meters longer than the altitude; find these dimensions.
$$b = 8·6 \text{ meters, } a = 5·4 \text{ meters. } Ans.$$

Hint. $\quad b = a + 3·2$ meters.
$$\therefore a(a + 3·2) = 46·44, \text{ etc.}$$

119. The perimeter of a rectangle is 13 meters longer than the base; the area is 20·88 square meters. Find the sides. $\quad$ 5·8 and 3·6 meters, or 7·2 and 2·9 meters. *Ans.*

120. The perimeter of a rectangle is 3·78 meters; its diagonal, 1·35. Find the area.
$$R = 0·8748 \text{ square meters. } Ans.$$

121. A rectangular field is 60 meters long by 40 meters wide. It is surrounded by a road of uniform width, the whole area of which is equal to the area of the field. Find the width of the road. $\quad$ 10 meters. *Ans.*

122. A rectangular court is 20 meters longer than broad, and its area is 4,524 square meters; find its length and breadth. $\quad$ 78 and 58 meters. *Ans.*

93. $y = b^2$.

123. Find the area of a square whose side is 15·4 meters.
$$237·16 \text{ square meters. } Ans.$$

124. The area of a square is $\tfrac{5}{8}$ square meter; find its side.
$$0·79057 \text{ meters. } Ans.$$

125. The side of a square is a; find the side when the area is n times as great. $\quad a_1 = \sqrt{n}$. *Ans.*

126. The sum of two squares is 900 square meters, the difference 252 square meters; find the sides.
$$24 \text{ and } 18 \text{ meters. } Ans.$$

Hint. $\qquad a_1^2 + a_2^2 = 909,$
$\qquad\qquad\quad a_1^2 + - a_2^2 = 252.$
$\qquad\qquad\quad \therefore a_1^2 = 576.$

127. The sides of two squares differ by 12 meters; their areas by 240 square meters. Find the side and area of each.   Sides, 4 and 16 meters;
Areas, 16 and 256 square meters. *Ans.*

HINT. $(a + 12)^2 - a^2 = 240.$

128. The perimeter of a square is 48 meters longer than the diagonal; find the area.
344·544375 square meters. *Ans.*

HINT. $4a = a\sqrt{2} + 48.$

129. The sum of the diagonal and side of a square is 100 meters; find the area.   1715·6164 square meters. *Ans.*

HINT. $2a^2 = (100 - a)^2.$

34. $\square = ab.$

130. The area of a parallelogram is 120 square meters, two sides are 12 and 14 meters; find both diagonals.
24 and 10·2 meters. *Ans.*

131. The altitudes a_1 and a_2 of a parallelogram are 5 and 8 meters; one diagonal is 10 meters. Find the area.
40·285 or 100·65 square meters. *Ans.*

HINT. $b_2 = \tfrac{5}{8}b_1,$
$b_1 = 10^2 + b_2^2 - 2b_2 j.$
Also, $j^2 = 100 - 64 = 36. \therefore j = 6.$

132. One side of a parallelogram is 8 meters longer than the corresponding altitude, and $\square = 384$ square meters; find this side.   24 meters. *Ans.*

35. $\triangle = \tfrac{1}{2}ab.$

133. The area of a rhombus is half the product of its diagonals.

134. From any point in an equilateral triangle the three perpendiculars on the sides together equal the altitude.

135. Find the area of a right triangle whose two sides are 248·2 and 160·5 meters.
19,918·05 square meters. *Ans.*

136. If 18·4 meters is the altitude of a $\triangle = 125{\cdot}36$ square meters, find b. 13·626 meters. *Ans.*

137. The two diagonals of a rhombus are 8·52 and 6·38 meters; find the area. 27·1788 square meters. *Ans.*

138. The altitude of a triangle is 8 meters longer than its base, and area is 44·02 square meters; find b.
6·2 meters. *Ans.*

Hint. $$\frac{b(b+8)}{2} = 44{\cdot}02.$$

139. The altitude of a right triangle cuts the hypothenuse into two parts, 7·2 and 16·2 meters long; find the area. 126·36 square meters. *Ans.*

36. $\triangle = \sqrt{s(s-a)(s-b)(s-c)}$.

140. $2\log\triangle = \log s + \log(s-a) + \log(s-b) + \log(s-c)$.

141. If b is the base of an equilateral triangle, find the area.

Here
$$a = \sqrt{b^2 - \frac{b^2}{4}} = \frac{b}{2}\sqrt{3}.$$
$$\therefore \triangle = ab = \frac{b^2}{4}\sqrt{3}.$$

142. The altitude of an equilateral triangle is 8·5 meters; find the area. $\triangle = \dfrac{a^2}{3}\sqrt{3} = 41{\cdot}71$ square meters. *Ans.*

143. The area of an equilateral triangle is 5·00548 square meters; find the side. 3·4. *Ans.*

EXERCISES AND PROBLEMS. 165

144. The side of an equilateral triangle is 4 meters longer than the altitude; find both.

$a = 25{\cdot}856, \; b = 29{\cdot}856.$ *Ans.*

HINT.
$$b^2 = (b-4)^2 + \frac{b^2}{4}.$$
$$b^2 - 32x = -64.$$

145. The three sides of a triangle are 10·2, 13·6, and 17 meters; find the area.   69·36 square meters. *Ans.*
Here
$s = 20{\cdot}4. \;\; \therefore \; \Delta = \sqrt{20{\cdot}4\,(20{\cdot}4 - 10{\cdot}2)\,(20{\cdot}4 - 13{\cdot}6)\,(20{\cdot}4 - 17)}.$

146. By measurement, $a = 37{\cdot}18$ meters, $b = 48{\cdot}72$ meters, $c = 56{\cdot}46$ meters; find Δ.
Here
$$
\begin{aligned}
s &= 71{\cdot}18; & \therefore \log s &= 1{\cdot}85236 \\
s - a &= 34{\cdot}00; & \therefore \log(s-a) &= 1{\cdot}53148 \\
s - b &= 22{\cdot}46; & \therefore \log(s-b) &= 1{\cdot}35141 \\
s - c &= 14{\cdot}72; & \therefore \log(s-c) &= 1{\cdot}16791 \\
& & \therefore \log \Delta^2 &= 5{\cdot}90316 \\
& & \therefore \log \Delta &= 2{\cdot}95158
\end{aligned}
$$
$\therefore \; \Delta = 894{\cdot}50$ square meters. *Ans.*

147. If three arcs, whose radii are 3, 2, 1, at their centers subtend angles of 60°, 90°, 120°, and intersect each other at their extremities, prove that the sides of the triangle formed by their chords are $3, 2\sqrt{2}, \sqrt{3}$; and its area $= \frac{1}{2}\sqrt{23}$.

HINT. The perpendicular from vertical $\angle$ 120° of isosceles Δ equals half a side, since joining its foot with midpoint of side makes an equilateral Δ.

148. The area of a triangle is 1012; the length of the side a is to that of b as 4 to 3, and c is to b as 3 to 2. Required the length of the sides.

$a = 52{\cdot}470, \; b = 39{\cdot}353, \; c = 59{\cdot}029.$ *Ans.*

Here
$$a = \tfrac{1}{3}b, \quad c = \tfrac{2}{3}b; \quad \therefore\ 2s = \tfrac{1}{3}b + b + \tfrac{2}{3}b = \tfrac{2+3+2}{3}b\ \tfrac{2\cdot 3}{6}b.$$
$$\therefore\ s\ \ = \tfrac{23}{12}b.$$
$$\therefore\ s - a = \tfrac{23-16}{12}b = \tfrac{7}{12}b,$$
and $\qquad s - b = \tfrac{11}{12}b,$
and $\qquad s - c = \tfrac{23-18}{12}b = \tfrac{5}{12}b.$
$$\therefore\ \text{Area} = \sqrt{\tfrac{23}{12} \cdot \tfrac{7}{12} \cdot \tfrac{11}{12} \cdot \tfrac{5}{12}b^4} = 1012.$$
$$\therefore\ \sqrt{8855}\,b^2 = 1012 \times 144 = 145{,}728.$$
$$\therefore\ 94{\cdot}101\,b^2 = 145{,}728.$$
$$\therefore\ b = \sqrt{\tfrac{145728}{94\cdot 101}} = 39{\cdot}353.$$

149. The area of a triangle is 144, and one of two equal sides is 24; find the third side, or base.

Here
$$s = 24 + \tfrac{b}{2}, \quad \text{and} \quad s - a = s - c = \tfrac{1}{2}b. \quad s - b = 24 - \tfrac{1}{2}b.$$
$$\therefore\ 144 = \sqrt{(24^2 - \tfrac{1}{4}b^2)\tfrac{1}{4}b^2}.$$
$$\therefore\ 576 = \sqrt{(2304 - b^2)\,b^2}.$$

150. Show that, in terms of its three medials,
$$\triangle = \tfrac{1}{3}\sqrt{2i_1^2 i_2^2 + 2i_2^2 i_3^2 + 2i_3^2 i_1^2 - i_1^4 - i_2^4 - i_3^4}.$$

Proof: $\qquad 4(i_1^2 + i_2^2 + i_3^2) = 3(a^2 + b^2 + c^2),$
$\qquad 16(i_1^2 i_2^2 + i_1^2 i_3^2 + i_2^2 i_3^2) = 9(a^2 b^2 + a^2 c^2 + b^2 c^2),$
$\qquad 16(i_1^4 + i_2^4 + i_3^4) = 9(a^4 + b^4 + c^4).$

But, by multiplying out 36, we have
$$\triangle = \tfrac{1}{4}\sqrt{2(a^2 b^2 + a^2 c^2 + b^2 c^2) - (a^4 + b^4 + c^4)}.$$

151. Prove that the triangle whose sides equal the medials of a given triangle is three-fourths of the latter.

Table II.—Scalene Triangles.

Sides.			Area.	Sides.			Area.
4	13	15	24	25	33	52	330
3	25	26	36	11	100	109	330
9	10	17	36	17	39	44	330
7	15	20	42	24	35	53	336
6	25	29	60	25	29	36	360
11	13	20	66	13	68	75	390
5	29	30	72	20	51	65	408
13	14	15	84	25	39	56	420
8	29	35	84	21	85	104	420
10	17	21	84	26	35	51	420
12	17	25	90	21			420
19	20	37	114	19	60	73	456
16	25	39	120	35	44	75	462
13	20	21	126	25	39	40	468
15	28	41	126	8	123	125	480
11	25	30	132	29	35	48	504
11	90	97	132	51	52	101	510
13	40	51	156	29	60	85	522
15	26	37	156	28	65	89	546
10	35	39	168	25	51	52	624
13	30	37	180	25	52	63	630
12	55	65	198	36	91	125	630
7	65	68	210	26	51	55	660
17	25	28	210	25	92	113	690
9	73	80	216	29	52	69	690
15	41	52	234	17	105	116	714
13	37	40	240	32	53	75	720
9	65	70	252	34	65	93	744
33	34	65	264	25	63	74	756
15	37	44	264	39	41	50	780
25	51	74	300	21	89	100	840
20	37	51	306	35	52	73	840

Table II.—*Continued.*

Sides.			Area.	Sides.			Area.
25	84	101	840	43	259	300	1,806
14	157	165	924	26	145	153	1,836
35	53	66	924	51	75	78	1,836
33	56	65	924	80	91	165	1,848
22	85	91	924	55	84	125	1,848
40	51	77	924	45	85	104	1,872
31	156	185	930	45	91	116	1,890
23	140	159	966	53	75	88	1,980
34	61	75	1,020	65	66	109	1,980
57	60	111	1,026	48	85	91	2,016
36	61	65	1,080	65	72	119	2,016
31	97	120	1,116	17	260	267	2,040
39	62	85	1,116	92	117	205	2,070
25	101	114	1,140	61	69	100	2,070
38	65	87	1,140	65	68	105	2,142
51	98	145	1,176	60	73	91	2,184
35	78	97	1,260	61	74	87	2,220
16	195	205	1,288	55	136	183	2,244
41	66	85	1,320	19	289	300	2,280
40	111	145	1,332	68	75	77	2,310
23	123	130	1,380	58	85	117	2,340
46	75	109	1,380	45	133	164	2,394
51	74	115	1,380	29	182	195	2,436
44	75	97	1,584	87	119	200	2,436
35	100	117	1,638	35	174	197	2,436
39	85	92	1,656	41	169	200	2,460
50	69	73	1,656	85	123	202	2,460
41	84	85	1,680	65	89	132	2,574
56	61	75	1,680	31	193	210	2,604
57	65	68	1,710	39	145	164	2,610
39	110	137	1,716	65	87	88	2,640
29	150	169	1,740	61	91	100	2,730
29	125	136	1,740	21	340	353	2,856
52	73	75	1,800	49	200	241	2,940

EXERCISES AND PROBLEMS.   169

TABLE II.—*Continued*.

Sides.			Area.	Sides.			Area.
27	275	292	2,970	105	124	205	5,208
35	197	216	3,024	75	176	229	5,280
76	85	105	3,196	51	233	260	5,304
37	195	212	3,330	65	173	204	5,304
87	112	185	3,360	45	296	325	5,328
45	164	187	3,366	91	125	174	5,460
78	95	97	3,420	104	111	175	5,460
57	122	125	3,420	99	113	140	5,544
65	109	116	3,480	47	250	267	5,640
73	102	145	3,480	55	244	273	6,006
65	126	173	3,484	105	116	143	6,006
65	119	156	3,570	100	217	303	6,510
40	231	257	3,696	91	145	180	6,552
69	113	140	3,864	153	185	328	6,660
65	119	138	3,864	43	520	555	6,708
60	145	161	3,864	119	150	241	7,140
89	99	100	3,960	50	369	401	7,380
57	148	175	3,990	89	170	189	7,560
75	109	136	4,080	65	297	340	7,722
85	99	140	4,158	37	525	548	7,770
91	100	159	4,200	85	234	293	7,956
90	97	119	4,284	123	133	200	7,980
40	291	325	4,290	65	272	303	8,160
87	100	143	4,290	111	200	281	8,880
68	87	145	4,350	140	143	157	9,240
39	280	305	4,368	68	273	275	9,240
89	111	170	4,440	111	175	176	9,240
55	207	244	4,554	89	208	231	9,240
66	175	221	4,620	116	231	325	9,240
143	168	305	4,620	111	175	232	9,324
61	155	156	4,650	74	277	315	9,324
37	411	440	4,884	117	164	175	9,450
41	337	360	4,904	116	181	225	10,440
123	208	325	4,920	91	253	300	10,626

TABLE II.—*Concluded.*

Sides.			Area.	Sides.			Area.
148	153	175	10,710	190	231	377	17,556
113	195	238	10,920	175	221	318	18,564
149	156	175	10,920	175	221	276	19,320
66	389	425	11,220	125	312	323	19,380
123	187	200	11,220	143	296	375	19,536
85	293	336	11,424	186	221	275	20,460
170	171	305	11,628	212	225	247	22,230
75	403	452	12,090	260	287	519	22,386
93	325	388	12,090	129	377	440	22,704
130	185	231	12,012	205	286	411	27,060
113	225	238	12,600	221	346	525	27,300
157	165	184	12,144	123	595	676	29,274
87	340	385	13,398	253	260	315	31,878
164	225	349	14,760	277	304	315	38,304
125	253	312	15,180	255	407	596	41,514
225	287	496	15,624	217	404	495	42,966
195	203	356	15,834	175	527	600	44,268
144	221	275	15,840	273	425	628	46,410
126	269	325	16,380				

37. $r = \dfrac{\Delta}{s}.$

152. Any right triangle equals the rectangle of the segments of the hypothenuse made by a perpendicular from center of inscribed circle.

153. In any triangle ABC, let M be the mid-point of the base AC, I the point of contact of the inscribed circle, H and K the points where the perpendicular from the vertex B, and the bisector of the angle B meet AC; prove the relation $MI \cdot HI = MH \cdot KI$.

EXERCISES AND PROBLEMS. 171

154. Each tangent from A equals $s-a$; from B equals $s-b$; from C equals $s-c$.

155. In a right triangle, CD is the perpendicular from C on hypothenuse; prove that the circles inscribed in triangles CAD, CBD have the same ratio as these triangles.

156. If h_1, h_2, h_3 be the perpendiculars from the angles of a triangle upon the opposite sides, and r the radius of the inscribed circle, prove $\dfrac{1}{h_1}+\dfrac{1}{h_2}+\dfrac{1}{h_3}=\dfrac{1}{r}$.

HINT. $\quad \dfrac{1}{h_1}=\dfrac{a}{2\Delta}\quad$ and $\quad \dfrac{1}{h_2}=\dfrac{b}{2\Delta}$, etc.

157. Prove $\quad h_1 h_2 h_3 = \dfrac{(a+b+c)^3 r^3}{abc}$.

158. If h'_1, h'_2, h'_3 be the perpendiculars from any point within a triangle, upon the sides, prove

$$\dfrac{h'_1}{h_1}+\dfrac{h'_2}{h_2}+\dfrac{h'_3}{h_3}=1.$$

159. If τ_1, τ_2, τ_3 be the distances from the angles of a triangle to the points of contact of the inscribed circle, prove

$$r = \left(\dfrac{\tau_1 \tau_2 \tau_3}{\tau_1+\tau_2+\tau_3}\right)^{\frac{1}{2}}.$$

160. If τ_4, τ_5, τ_6 be the distances from the angles of a triangle to the center of the inscribed circle, prove

$$r = \tfrac{1}{2}\dfrac{\tau_4 \tau_5 \tau_6}{abc}(a+b+c).$$

161. Prove $\quad abc = a\tau_4^2 + b\tau_5^2 + c\tau_6^2$.

162. Prove $\tau_4^2 + \tau_5^2 + \tau_6^2 = ab+ac+bc - \dfrac{6abc}{a+b+c}$.

38. $\mathfrak{R} = \dfrac{abc}{4\Delta}$.

163. The radius of a circle is 8 meters. Find the side of an inscribed equilateral triangle. $b = 13{\cdot}8564$ meters. *Ans.*

164. Find the radius circumscribing the equilateral triangle whose base equals 8·66 meters.   5 meters. *Ans.*

165. In every triangle, the sum of the perpendiculars from the center of the circumscribed circle on the three sides is equal to the sum of the radii of the inscribed and circumscribed circles.

166. If from the vertices of an equilateral triangle perpendiculars be drawn to any diameter of the circle circumscribing it, the perpendicular which falls on one side of this diameter will be equal to the sum of the two which fall on the other side.

167. If the altitude of an isosceles triangle is equal to its base, $\frac{5}{8}b = \mathfrak{R}$.

168. If A', B', C' be the feet of the perpendiculars from the angles of a triangle upon the sides, prove that the radius circumscribing ABC is twice the radius circumscribing $A'B'C'$.

169. If a, β, γ be the perpendiculars from the center of the circumscribing circle upon a, b, c, the sides of a triangle, prove
$$\frac{abc}{a\beta\gamma} = 4\left(\frac{a}{a} + \frac{b}{\beta} + \frac{c}{\gamma}\right).$$

39. $r_1 = \dfrac{\Delta}{s-a}$; $r_2 = \dfrac{\Delta}{s-b}$; $r_3 = \dfrac{\Delta}{s-c}$.

170. If the sides of a triangle be in arithmetical progression, the perpendicular on the mean side from the opposite angle, and the radius of the circle which touches the mean side and the two other sides produced, are each three times the radius of the inscribed circle.

171. Each of the common outer tangents to two circles equals the part of the common inner tangent intercepted between them.

172. Each tangent from A to the circle escribed to a equals s; from B to circle escribed to a equals $s-c$; from C to circle escribed to a equals $s-b$. Similar theorems hold for the escribed circles which touch b and c.

173. The area of a triangle of which the centers of the escribed circles are the angular points is $\dfrac{abc}{2r}$.

174. If a, b, c denote the sides of a triangle; h_1, h_2, h_3 the three altitudes; q_1, q_2, q_3 the sides of the three inscribed squares, prove the relations

$$\frac{1}{q_1}=\frac{1}{h_1}+\frac{1}{a}; \quad \frac{1}{q_2}=\frac{1}{h_2}+\frac{1}{b}; \quad \frac{1}{q_3}=\frac{1}{h_3}+\frac{1}{c}.$$

175. Prove
$$\frac{1}{r_1}=-\frac{1}{h_1}+\frac{1}{h_2}+\frac{1}{h_3};$$
$$\frac{1}{r_2}=+\frac{1}{h_1}-\frac{1}{h_2}+\frac{1}{h_3}; \text{ etc.}$$

176. Prove
$$\frac{2}{h_1}=\frac{1}{r}-\frac{1}{r_1}=\frac{1}{r_2}+\frac{1}{r_3};$$
$$\frac{2}{h_2}=\frac{1}{r}-\frac{1}{r_2}=\frac{1}{r_3}+\frac{1}{r_1};$$
$$\frac{2}{h_3}=\frac{1}{r}-\frac{1}{r_3}=\frac{1}{r_1}+\frac{1}{r_2}.$$

177. Prove $\quad \tfrac{1}{2}h_1=\dfrac{rr_1}{r_1-r}=\dfrac{r_2 r_3}{r_2+r_3}.$

178. If $\tau, \tau_7, \tau_8, \tau_9$ are the distances from the center of the circumscribed circle to the centers of the inscribed and escribed circles, prove the relations

$$\mathfrak{R}^2 = \tau^2 + 2r\mathfrak{R} = \tau_7^2 - 2r_1\mathfrak{R} = \tau_8^2 - 2r_2\mathfrak{R}$$
$$= \tau_9^2 - 2r_3\mathfrak{R} = \frac{\tau^2 + \tau_7^2 + \tau_8^2 + \tau_9^2}{12}.$$

HINT. $\quad \tau^2 = \mathfrak{R}^2 - 2r\mathfrak{R}; \quad \tau_7^2 = \mathfrak{R}^2 + 2r_1\mathfrak{R};$
$\quad\quad\quad \tau_8^2 = \mathfrak{R}^2 + 2r_2\mathfrak{R}; \quad \tau_9^2 = \mathfrak{R}^2 + 2r_3\mathfrak{R}.$

179. Prove $\quad r\mathfrak{R} = \dfrac{abc}{2(a+b+c)}$;

$r_1\mathfrak{R} = \dfrac{abc}{4(s-a)}$; $\quad r_2\mathfrak{R} = \dfrac{abc}{4(s-b)}$; $\quad r_3\mathfrak{R} = \dfrac{abc}{4(s-c)}$;

180. Prove $\quad r_1 + r_2 + r_3 = r + 4\mathfrak{R}$.

181. In any triangle prove $s^2 - i_1^2 = rr_1$; etc.

40. $T = x\dfrac{y_1 + y_2}{2}$.

182. The base of a triangle is 20 meters, and its altitude 18 meters. It is required to draw a line parallel to the base so as to cut off a trapezoid containing 80 square meters. What is the length of the line of section, and its distance from the base of the triangle?

Calling b_2 the line of section, and x its distance from b_1,

$$T = 80 = \tfrac{1}{2}(20 + b_2)x.$$

Now, $\quad b_2 : 20 :: 18 + x : 18.$

$\therefore\ b_2 = \tfrac{20}{18}(18-x) = \tfrac{10}{9}(18-x) = 20 - \tfrac{10}{9}x.$
$\therefore\ 80 = \tfrac{1}{2}(20 + 20 - \tfrac{10}{9}x)x = (20 - \tfrac{5}{9}x)x.$
$\therefore\ 720 = 180x - 5x^2.$

$\therefore\ x^2 - 36x = -144.$
$\therefore\ x - 18 = \sqrt{324 - 144}.$
$\therefore\ x = 18 - \sqrt{180}.$

$\therefore\ x = 18 - 13\cdot 416 = 4\cdot 584,$
and $\quad b_2 = 20 - \tfrac{10}{9}(4\cdot 584) = 20 - 5\cdot 093 = 14\cdot 907.$

183. In a perpendicular section of a ditch, the breadth at the top is 26 feet, the slopes of the sides are each 45°, and the area 140 square feet. Required the breadth at bottom and the depth of the ditch.

Here

$T = 140 = \frac{1}{2}[26 + 26 - 2x]x = [26 - x]x.$
$\therefore 140 = 26x - x^2.$
$\therefore x = 13 - \sqrt{169 - 140} = 13 - \sqrt{29} = 13 - 5\cdot 385 = 7\cdot 615.$
$\therefore b_1 = 26 - 15\cdot 230 = 10\cdot 77.$

184. The altitude of a trapezoid is 23 meters; the two parallel sides are 76 and 36 meters; it is required to draw a line parallel to the parallel sides, so as to cut off from the smaller end of the trapezoid a part containing 560 square meters. What is the length of the line of section, and its distance from the shorter of the two parallel sides?

Let x equal altitude of required part.

$T_1 = \frac{1}{2}[76 + 36]23 = 1288,$
$T_2 = 728 = \frac{1}{2}[76 + l][23 + x].$

$\therefore \dfrac{1456}{76 + l} = 23 - x. \quad \therefore x = 23 - \dfrac{1456}{76 + l}.$

Also, $\quad T_3 = 560 = \frac{1}{2}[36 + l]x.$

$\therefore \dfrac{1120}{36 + l} = x.$

$\therefore \dfrac{1120}{36 + l} = 23 - \dfrac{1456}{76 + l}.$

$\therefore 137{,}536 + 2576 \cdot l = 62928 + 2576 \cdot l + 23 \cdot l^2,$
$\underline{\phantom{137{,}536\ } 62{,}928}$
$74{,}608 = 23 \cdot l^2.$

$\therefore l^2 = 3243\cdot 8.$
$\therefore l = 56\cdot 95.$

$x = \dfrac{1120}{36 + 56\cdot 95} = 12\cdot 048.$

185. The two parallel sides of a trapezoid are 83·2 and 110·4 meters; the altitude, 50·4 meters. Find the area.
$T = 5227\cdot 2$ square meters. *Ans.*

186. The perimeter of a trapezoid is 122 meters. The non-parallel sides are 36 and 32 meters; the altitude, 30·4 meters. Find the area. $T = 820\cdot 8$ square meters. *Ans.*

187. $T = 151\cdot9$ square meters, $a = 12\cdot4$ meters, $b_1 = 18\cdot6$ meters. Find the other parallel side.

$b_2 = 5\cdot9$ meters. *Ans.*

188. The altitude and two parallel sides of a trapezoid are $2:3:5$, and $T = 1270\cdot08$ square meters. Find the parallel sides. $b_1 = 63$ meters; $b_2 = 37\cdot8$ meters. *Ans.*

189. The triangle formed by joining the mid-point of one of the non-parallel sides of a trapezoid to the extremities of the opposite side is equivalent to half the trapezoid.

190. The area of a trapezoid is equal to half the product of one of its non-parallel sides, and the perpendicular from the mid-point of the other upon the first.

191. The line which joins the mid-points of the diagonals of a trapezoid is parallel to the bases, and equals half their difference.

192. Cutting each base of a trapezoid into the same number of equal parts, and joining the corresponding points, divides the trapezoid into that number of equivalent parts.

193. If the mean line of a trapezoid be divided into n equal parts, and through these points lines, not intersecting within the trapezoid, be extended to its bases, they cut the trapezoid into n equal trapezoids.

194. In every trapezoid, the difference of the squares of the diagonals has to the difference of the squares of the non-parallel sides the same ratio that the sum of the parallel sides has to their difference.

195. Let b_1 be the longer, b_2 the shorter, of the two parallel sides in any trapezoid, z_1 and z_2 the other two sides, and take

$$A = \sqrt{(b_1 - b_2 + z_1 + z_2)(b_2 - b_1 + z_1 + z_2)(b_1 - b_2 - z_1 + z_2)(b_1 - b_2 + z_1 - z_2)}.$$

EXERCISES AND PROBLEMS. 177

Prove $\quad T = \dfrac{b_1 + b_2}{4(b_1 - b_2)} \Lambda.$

From the intersection point of b_2 and z_2 draw a line parallel to z_1; the base of the triangle so formed is $(b_1 - b_2)$, and its other sides are z_1 and z_2.

∴ by 36,
$$\Delta = \tfrac{1}{4} \Lambda.$$
$$\therefore a = \dfrac{\tfrac{1}{4} \Lambda}{\tfrac{1}{2}(b_1 - b_2)} = \dfrac{\Lambda}{2(b_1 - b_2)}.$$
$$\therefore T = \tfrac{1}{2} a (b_1 + b_2) = \dfrac{b_1 + b_2}{4(b_1 - b_2)} \Lambda.$$

196. The two parallel sides of a trapezoid are 184 and 68 meters; the two others, 84 and 72 meters. Find the area.   6536 square meters. *Ans.*

197. The diagonal of a symmetric trapezoid is $\sqrt{z^2 + b_1 b_2}$.

198. The altitude of a trapezoid is 80 meters; the two diagonals 110 and 100 meters. Find the area.
     5419·6 square meters. *Ans.*

HINT. $\quad T = \tfrac{1}{2} a (b_1 + b_2) = \tfrac{1}{2} a (\sqrt{c_1^2 - a^2} + \sqrt{c_2^2 - a^2}).$

199. In a trapezoid $a = 140$, $b_1 = 160$, $b_2 = 120$ meters; if the area is halved by a line parallel to the bases, find its length and distance below the shorter base.
     $l = 141 \cdot 42$ meters, $d = 74 \cdot 97$ meters. *Ans.*

HINT I. $\quad\quad\quad l - b_2 : b_1 - b_2 :: d : a,$
   II. $\quad \dfrac{l + b_2}{2} d = \dfrac{T}{2} = \dfrac{b_1 + b_2}{4} a = 9800;$

or,   I. $\quad\quad\quad\quad 7l - 2d = 840,$
   II. $\quad\quad\quad 120 d + ld = 19{,}600.$

Substitute the value of d from I. in II.

200. In a trapezoid $b_1 = 312$, $b_2 = 39$, $z_1 = 350$, $z_2 = 287$ meters; if cut by parallels to b into three similar trapezoids, find where the two parallels cut the sides, and find the areas of the three trapezoids.

If $\dfrac{l_2}{b_2} = n,$ then $l_2 = nb_2.$

$\therefore l_1 = nl_2 = n^2 b_2.$

$\therefore b_1 = nl_1 = n^2 l_2 = n^3 b_2 = 39\,n^3 = 312.$

$\therefore n = 2.$

$\therefore z_3 : z_5 : z_7 :: l_2 : l_1 : b_1 :: 1 : 2 : 4.$

$\therefore z_3 = \dfrac{z_1}{7} = 50$ meters.

$z_4 = \dfrac{z_2}{7} = 41$ meters, etc.

41. $\sum\limits_{v=1}^{v=n} T_v = \tfrac{1}{2}[(x_2 - x_1)(y_1 - y_3) + (x_3 - x_1)(y_2 - y_4) + \cdots$
$+ (x_n - x_1)(y_{n-1} - y_{n+1}) + (x_{n+1} - x_1)(y_n + y_{n+1})].$

201. Find the distance between the points 1 and 2. Between two points $(x_1 y_1)$, $(x_2 y_2)$ the distance

$$r = \sqrt{(x_2 - x_1)^2 + (y_2 - y_1)^2}.$$

202. Find the sum of the two right trapezoids determined by the ordinates of the three points (12·3, 45·6), (78·9, 13), (24, 57).

203. If the cross section of an excavation is a trapezoid, b breadth of top, h depth, with side slopes m and n in 1, which means that one side falls m meters vertically for one meter of horizontal distance; then show $T = bh - \dfrac{m+n}{2\,mn} h^2.$

42. $N = \tfrac{1}{2}[x_1(y_n - y_2) + x_2(y_1 - y_3) + x_3(y_2 - y_4) + \cdots$
$+ x_n(y_{n-1} - y_1)].$

204. Prove that a polygon may be constructed when all but three adjacent parts (1 side and 2 ∡s, or 2 sides and 1 ∡) are given. What theorem for congruence of polygons follows from this?

EXERCISES AND PROBLEMS. 179

205. Find the area of a heptagon from the coördinates of its vertices, measured as follows:

	x	y
1	0	1·72
2	10·48	16·84
3	16·26	14·36
4	32·54	4·84
5	50·02	10·32
6	50·02	0
7	0	0

$N = 476·21$ square meters. *Ans.*

206. Find the area of an enneagon from the following measurements.

	x	y
1	0	16·96
2	26·36	20·04
3	58·02	22·16
4	104·00	11·24
5	97·48	2·48
6	92·22	−11·86
7	61·00	− 2·36
8	35·46	− 4·10
9	9·84	−14·22

Also draw the figure.

$N = 2429·16$ square meters. *Ans.*

207. Find the area of a pentagon, the coördinates of whose vertices are as follows: (133, 917), (261, 325), (486, 916), (547, 325), (828, 916).

MODEL SOLUTION.

208.

Vertices.	Ordinates. y_m. ± Meters.	Ordinates. Difference. $y_{m-1}-y_{m+1}$. ± Meters.	Abscissae. ± Meters.	Abscissae. Difference. $x_{m+1}-x_{m-1}$. ± Meters.	Double Area. $y_m(x_{m+1}-x_{m-1})$. +Sq. Meters.	Double Area. $y_m(x_{m+1}-x_{m-1})$. −Sq. Meters.	Double Area. $x_m(y_{m-1}-y_{m+1})$. +Sq. Meters.	Double Area. $x_m(y_{m-1}-y_{m+1})$. −Sq. Meters.
1	0	− 837·064	0	−1,080·911				
2	+ 837·064	−1,386·464	− 183·722	+ 766·893	641,938·6		254,793·3	88,540·9
3	+1,386·464	− 115·454	+ 766·893	+1,598·809	2,216,690·0			73,683·2
4	+ 952·518	− 52·073	+1,415·037	+1,435·171	1,367,025·6			
5	+1,438·537	+ 182·240	+2,202·064	+1,452·368	2,089,277·1		401,304·1	
6	+ 770·278	+1,268·218	+2,867·405	+ 493·805	380,367·1		3,636,495·3	
7	+ 170·319	+ 339·356	+2,695·869	− 817·984		139,285·8	914,859·6	
8	+ 430·922	+ 549·886	+2,049·421	− 884·471		381,569·2	1,126,949·6	
9	− 379·567	+ 955·946	+1,810·398	− 841·719	319,488·7		1,730,642·1	
10	− 525·024	− 379·567	+1,207·702	− 913·259	479,482·7			458,401·8
11	0	− 525·024	+ 897·139	−1,207·702				471,005
					7,494,270	520,855	8,065,044	1,091,631
					520,855		1,091,631	
					6,973,415 = double area = 6,973,413			

Therefore, the area of the Hendecagon is 3,486,707 square meters.

209. Find the area of a hexagon from its coördinates (719, 313), (512, 852), (719, 454), (513, 116), (720, 242), (513, 993).

43. $2Q = x_1y_2 - x_2y_1 + x_2y_3 - x_3y_2 + x_3y_4 - x_4y_3 + x_4y_1 - x_1y_4$.

210. The area of a quadrilateral inscribed in a circle is
$= \sqrt{(s-a)(s-b)(s-c)(s-d)}$ where $s = \dfrac{a+b+c+d}{2} = \tfrac{1}{2}p$.

211. If through the mid-point E of the diagonal BD of a quadrilateral $ABCD$, FEG be drawn parallel to the other diagonal AC, prove that the straight line AG divides the quadrilateral into two equivalent parts.

212. Show that two quadrilaterals whose diagonals contain the same angle are as the products of their diagonals.

213. A circle of r is inscribed in a kite, and another of r' in the triangle formed by the axis of the kite and the two unequal sides; show that, if $2l$ be the length of the other kite-diagonal,
$$\frac{1}{r'} - \frac{1}{r} = \frac{1}{l}.$$

214. To find the area of any quadrilateral from one side, and the distances from that side of the other two vertices, and the intersection-point of the diagonals.

Given the side $AB = b$, and the ordinates from C, D, and E; namely, y_3, y_4, and y_5.

Parallel to BD draw CM to intersection with AD prolonged, and drop y_6, the ordinate of M.

Then $\quad\quad ABCD = \triangle AMB = \tfrac{1}{2}by_6$.

But $\quad\quad y_6 : y_4 = AM : AD = AC : AE = y_3 : y_5;$

$\therefore y_6 = \dfrac{y_3 y_4}{y_5};\quad\quad \therefore Q = \dfrac{b y_3 y_4}{2 y_5}.$

44. $A_1 = \dfrac{A_2 a_1^2}{a_2^2}.$

215. The area of a triangle equals 3259·6 square meters; one side equals 112·4 meters. Find the area of a similar triangle whose corresponding side equals 28·1 meters.
203·725 square meters. *Ans.*

216. The sides of a triangle are 389·2, 486·5, and 291·9 meters. The area of a similar triangle is 2098·14 square meters. Find its sides.   74·8, 56·1, 93·5 meters. *Ans.*

217. The areas of two similar triangles are 24·36 and 182·7 square meters. One side of the first is 8·5 meters shorter than the homologue of the second. Find these sides.   4·88 and 13·38 meters. *Ans.*

HINT. $24·36 : 182·7 :: (x - 8·5)^2 : x^2.$

218. Two triangles are 21·66 and 43·74 square meters, and have an equal angle whose including sides in the first are 7·6 and 5·7 meters. The corresponding sides in second differ by 2·7 meters. Find them.
10·8 and 8·1 meters. *Ans.*

219. The areas of two similar polygons are 46·37 and 185·48 square meters. A side of the first is 15 meters smaller than the corresponding side of the other. Find these sides.   15 and 30 meters. *Ans.*

HINT. $46·37 : 185·48 : x^2 : (x + 15)^2.$

45. $N = \dfrac{aln}{2} = \dfrac{ap}{2}.$

220. The sum of perpendiculars dropped from any point within a regular polygon upon all the sides is constant.

221. In area, an inscribed $2n$-gon is a mean proportional between the inscribed and circumscribed n-gons.

222. With what regular polygons can a vestibule be paved?

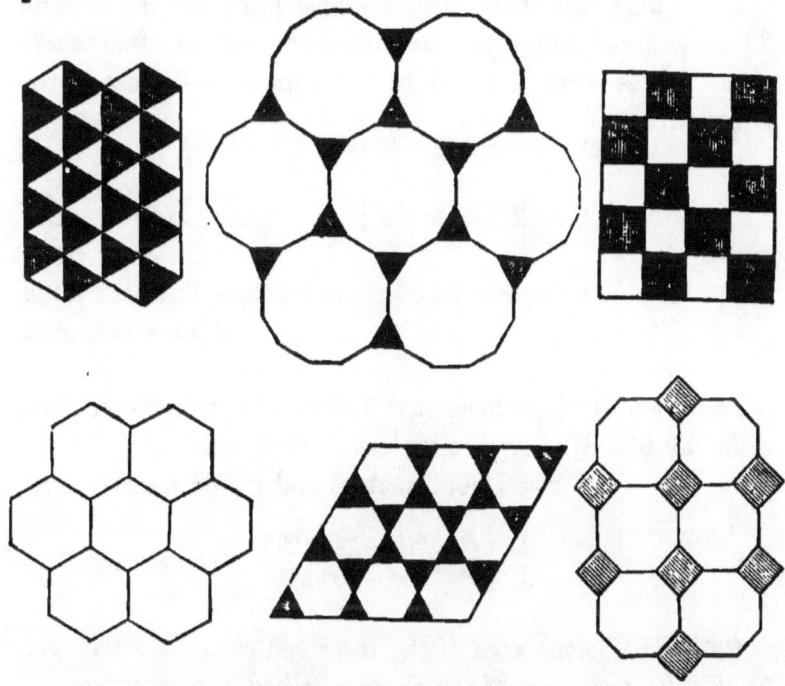

223. If a regular n-gon is revolved about its center through the $\angle \dfrac{\delta}{n}$, it coincides with itself.

224. In a regular n-gon, each $\angle = f - \dfrac{\delta}{n}$.

46. $N_l = l_n^2 N_1$.

225. A hexagon is inscribed in a circle, and the alternate angles are joined, forming another hexagon. Find its area.

$\dfrac{\sqrt{3}}{2} r^2$. *Ans.*

226. What is the area of a regular dodecagon whose side is 54 feet?

$(54)^2 = 2916$, and $2916 \times 11\cdot 1961524 = 32647\cdot 980+$. *Ans.*

47. $\odot = r^2\pi.$

227. There are three circles whose radii are 20, 28, and 29 meters respectively. Required the radius of a fourth circle whose area is equal to the sum of the areas of the other three.
$$\odot_1 = 400\pi, \quad \odot_2 = 784\pi, \quad \odot_3 = 841\pi;$$
$$\therefore \odot_4 = 2025\pi = r^2\pi; \therefore r = \sqrt{2025} = 45. \text{ } Ans.$$

228. If a circle equals 34·36 square meters, find its radius.
3·3 meters. *Ans.*

229. Two ⊙'s together equal 740·4232 square meters, and differ by 683·8744 square meters. Find radii.
$r = 15{\cdot}056$ meters and $r' = 3$ meters. *Ans.*

I. $\pi r^2 + \pi r'^2 = 740{\cdot}4232.$
II. $\pi r^2 - \pi' r'^2 = 683{\cdot}8744.$

230. If ⊙ be the area of the inscribed circle of a triangle, $\odot_1$, $\odot_2$, $\odot_3$ the areas of the three escribed circles, prove
$$\frac{1}{\sqrt{\odot_1}} + \frac{1}{\sqrt{\odot_2}} + \frac{1}{\sqrt{\odot_3}} = \frac{1}{\sqrt{\odot}}.$$

231. If from any point in a semicircumference a perpendicular be dropped to the diameter, and semicircles described on these segments, the area between the three semicircumferences equals the circle on the perpendicular as diameter.

232. The perimeters of a circle, a square, and an equilateral triangle are each of them 12 meters. Find the area of each of these figures to the nearest hundredth of a square meter.   11·46, 9, 6.93 square meters. *Ans.*

233. Find the side of a square inscribed in a semicircle.
$\frac{2}{5} r \sqrt{5}$. *Ans.*

234. An equilateral triangle and a regular hexagon have the same perimeter; show that the areas of their inscribed circles are as 4 to 9.

235. How far must the diameter of a circle be prolonged, in order that the tangent to the circle from the end of the prolongation may be m long? $\frac{1}{2}(\sqrt{d^2+4m^2}-d)$. *Ans.*

48. $S = \frac{1}{2} lr = \frac{1}{2} ur^2$.

236. Find the area of a sector of 68° 36' when $r = 7\cdot 2$.
31·03398 square meters. *Ans.*

237. When circle equals 432 square meters, find sector of 84° 12'.   100·8 square meters. *Ans.*

Hint. $432 : S :: 360 : 84\frac{1}{5}$.

238. Find the number of degrees in the arc of a sector equivalent to the square of its radius.

239. In different circles, sectors are equivalent whose angles have a ratio inverse to that of the squared radii.

240. Find radius when sector of 7° 12' is 2 square centimeters.

241. Find sector whose radius equals 25, and the circular measure of whose angle equals $\frac{3}{4}$.   234·375. *Ans.*

242. The length of the arc of a sector of a given circle is 16 meters; the angle of the sector at center is $\frac{1}{4}$ of a right angle. Find sector.   488·9 square meters. *Ans.*

49. $G = \dfrac{h^2(l+k) + \frac{1}{4}k^2(l-k)}{4h}$.

243. AB is a chord of a given circle; if on the radius CA, which passes through one of its extremities, taken as diameter, a circle be described, the segments cut off from the two circles by the chord AB are in the ratio of 4 to 1.

244. Show that, if a is the angle or arc of a segment, for

$a = 60°$, $\quad G = \dfrac{r^2}{12}(2\pi - 3\sqrt{3})$;

$a = 120°$, $\quad G = \dfrac{r^2}{12}(4\pi - 3\sqrt{3})$;

$a = 90°$, $\quad G = \dfrac{r^2}{\pi}(\pi - 2)$;

$a = 36°$, $\quad G = \dfrac{r^2}{40}(4\pi - 5\sqrt{10 - 2\sqrt{5}})$;

$a = 72°$, $\quad G = \dfrac{r^2}{40}(8\pi - 5\sqrt{10 - 2\sqrt{5}})$.

245. In a segment of 60°, to how many places of decimals is our approximation correct?

246. Prove that there can be no segment with $k = 120$, $l = 156$, $h = 12$.

247. In a circle, given two parallel chords k_1 and k_2, and their distance apart τ; find the diameter.

$$d = \sqrt{\dfrac{k_1^2 - k_2^2}{4\tau} + \tfrac{1}{2}(k_1^2 + k_2^2 + 2\tau^2)}. \quad Ans.$$

HINT. If $x = \sqrt{r^2 - \tfrac{1}{4}k_2^2}$ and $y = \sqrt{r^2 - \tfrac{1}{4}k_1^2}$,
then
  I. $x - y = \tau$.
  II. $x^2 - y^2 = \tfrac{1}{4}(k_1^2 - k_2^2)$.

50.

248. What is the area of a circular zone, one side of which is 96 and the other 60, and the distance between them 26 ($r = 50$), when the area of the larger sector is 3217·484, and of the smaller 1608·736?   2136·75. *Ans.*

51.

249. HIPPOCRATES'S THEOREM.

The two crescents made by describing semicircles outward on the two sides of a right triangle, and a semicircle toward them on the hypothenuse, are equivalent to the right triangle.

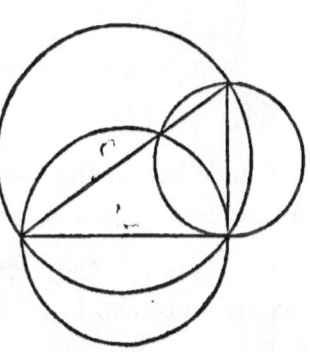

250. The crescent made by describing a semicircle on the chord of a quadrant equals the right triangle.

52. $A = (r_1 + r_2)(r_1 - r_2)\pi.$

251. A circle of 60 meters diameter is divided into seven equal parts by concentric circles; find the parts of the diameter.

$$r^2\pi = 900 \times 3\cdot 14159 = 2827\cdot 431.$$

$$\therefore \text{ outer annulus} = \frac{2827\cdot 431}{7} = 403\cdot 9186 = (900 - r_2^2)\pi.$$

$$\therefore r_2^2 = 900 - \frac{403\cdot 9186}{3\cdot 14159} = 900 - 128\cdot 575 = 771\cdot 425.$$

$$\therefore r_2 = 27\cdot 77+.$$

$$\therefore d_2 = 55\cdot 55.$$

$$\therefore \text{ 1st part} = 60 - 55\cdot 55 = 4\cdot 450+.$$

In the same way, 2d part = 4·840,
3d part = 5·353+,
4th part = 6·076+, etc.

252. Find the annulus between the concentric circumferences, $c_1 = 21{\cdot}98$ meters and $c_2 = 18{\cdot}84$ meters, taking $\pi = 3{\cdot}14$. $A = 10{\cdot}205$ square meters. *Ans.*

53. S. $A = \tfrac{1}{2} h (l_1 + l_2)$.

253. To trisect a sector of an annulus by concentric circles.

54. $J = \tfrac{2}{3} hk$.

254. What is the area of a parabola whose base is 18 meters and height 5 meters?   60 square meters. *Ans.*

255. What is the area of a parabola whose base is 525 meters and height 350 meters?
$\qquad\qquad\qquad\qquad$ 122,500 square meters. *Ans.*

55. $E = ab\pi$.

256. The area of an ellipse is to the area of the circumscribed circle as the minor axis is to the major axis.

257. The area of an ellipse is to the area of the inscribed circle as the major axis is to the minor axis.

258. The area of an ellipse is a mean proportional between the inscribed and circumscribed circles.

259. What is the area of an ellipse whose major axis is 70 meters, and minor axis 60 meters?
$\qquad\qquad\qquad\qquad$ 3298·67 square meters. *Ans.*

260. What is the area of an ellipse whose axes are 340 and 310?   82,780·896. *Ans.*

Exercises and Problems on Chapter IV.

Polyhedrons.

56. $\mathfrak{F} + \mathfrak{S} = \mathfrak{E} + 2.$

261. The number of plane angles in the surface of any polyhedron is twice the number of its edges.

HINT. Each face has as many plane angles as sides. Each edge pertains, as side, to two faces.

262. The number of plane angles on the surface of a polyhedron is always an even number.

263. If a polyhedron has for faces only polygons with an odd number of sides, *e.g.*, trigons, pentagons, heptagons, etc., it must have an even number of faces.

264. If the faces of a polyhedron are partly of an even, partly of an odd number of sides, there must be an even number of odd-sided faces.

265. In every polyhedron $\tfrac{3}{2}\mathfrak{F} \leqslant \mathfrak{E}$.

HINT. The number of plane angles on a polyhedron can never be less than thrice the number of faces.

266. In every polyhedron $\tfrac{3}{2}\mathfrak{S} \leqslant \mathfrak{E}$.

267. In any polyhedron $\mathfrak{E} + 6 \leqslant 3\mathfrak{S}$.

268. In any polyhedron $\mathfrak{E} + 6 \leqslant 3\mathfrak{F}$.

269. In every polyhedron $\mathfrak{E} < 3\mathfrak{S}$, and $\mathfrak{E} < 3\mathfrak{F}$.

270. In a polyhedron not all the summits are more than five-sided; nor have all the faces more than five sides.

271. There is no seven-edged polyhedron.

272. For every polyhedron $\hat{s} = \delta(\mathfrak{E} - \mathfrak{F})$, that is, the sum of the plane angles is as many perigons as the difference between the number of edges and faces.

273. For every polyhedron $\hat{s} = \delta(\mathfrak{S} - 2)$, just as for every polygon $\hat{s} = f(n - 2)$.

274. How many regular convex polyhedrons are possible?

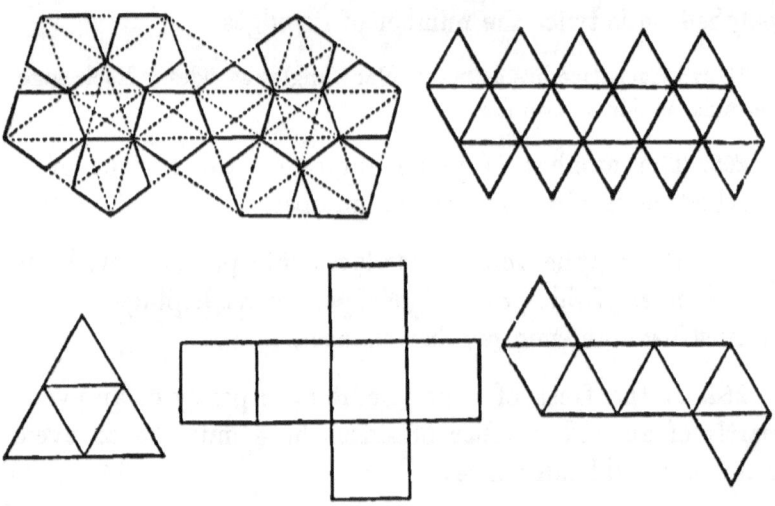

275. In no polyhedron can triangles and three-faced summits both be absent; together are present at least eight.

276. A polyhedron without triangular and quadrangular faces has at least twelve pentagons; a polyhedron without three-faced and four-faced summits has at least twelve five-faced.

57. $P = lp$.

277. In a right prism of 9 meters altitude, the base is a right triangle whose legs are 3 and 4 meters. Find the mantel.

278. The base of a right prism 12 meters tall is a triangle whose sides are 12, 14, and 15 meters. Find its surface.

279. To find the mantel of a truncated prism.
Rule: Multiply each side of the perimeter of the right section by the sum of the two edges in which it terminates. The sum of these products will be twice the area.

280. The mantel of a truncated prism equals the axis multiplied by perimeter of a right section.

281. A right prism 4 meters tall has for base a regular hexagon whose side is 1·2 meters. Find its surface.

282. In a right triangular prism the lateral edges equal the radius of the circle inscribed in the base. Show that the mantel equals the sum of the bases.

58. $C = cl = 2\pi r l$.

283. In a right circular cylinder,

(1) Given a and C; find r. $\dfrac{C}{2a\pi}$. Ans.
(2) Given B and C; find a.
(3) Given C and $a = 2r$; find surface.
(4) Given surface and $a = r$; find C.
(5) Given a and $B + C$; find r.

284. The mantel of a right cylinder is equal to a circle whose radius is a mean proportional between the altitude of the cylinder and the diameter of its base.

285. The bases of a circular cylinder together are to the mantel as radius to altitude.

286. If the altitude of a right circular cylinder is equal to the diameter of its base, the mantel is four times the base.

287. Find a cylinder equivalent to a sum of right circular cylinders of the same height.

HINT. Find a radius whose square equals squares of the n radii.

288. How much must the altitude of a right circular cylinder be prolonged to make its mantel equal its previous surface?

289. A plane perpendicular to the base of a right cylinder cuts it in a chord whose angle at the center is a; find the ratio of the curved surfaces of the two parts of the cylinder.

59. $Y = \frac{1}{2} hp$.

290. The surface of any regular tetrahedron is to that of the cube on its edge as 1 to $2 \cdot \sqrt{3}$.

291. Each edge of a regular tetrahedron is 2 meters. Find mantel.

292. Each edge of a regular square pyramid is 2 meters. Find surface.

293. From the altitude a and basal edge b of a regular hexagonal pyramid, find its surface.
$$3b(\tfrac{1}{2}b\sqrt{3} + \sqrt{a^2 + \tfrac{3}{4}a^2}). \quad Ans.$$

294. In a regular square pyramid, given p, the perimeter of the base, and the area $\triangle$ of the triangle made by a basal diagonal and the two opposite lateral edges; find the surface of the pyramid. $\quad \tfrac{1}{16}p^2 + \tfrac{1}{2}\sqrt{32\triangle^2 + \tfrac{1}{64}p^4}. \quad Ans.$

60. $K = \frac{1}{2} ch = \pi rh$.

295. The convex surface of a right cone is twice the area of the base; find the vertical angle.

Here
$$c \times \tfrac{1}{2} h = 2c \times \tfrac{1}{2} r.$$
$$\therefore h = 2r = d.$$

EXERCISES AND PROBLEMS.

Thus the section containing the axis is an equilateral triangle; so the angle equals 60°. *Ans.*

296. Find the ratio of the mantels of a cone and cylinder whose axis-sections are equilateral.

297. Find the locus of the point equally distant from three given points.

298. In a right cone

(1) Given a and r; find K. $\quad r\pi\sqrt{a^2+r^2}$. *Ans.*

(2) Given a and h; find K. $\quad \pi h\sqrt{h^2-a^2}$. *Ans.*

(3) Given K and h; find r. $\quad \dfrac{K}{\pi h}$. *Ans.*

299. In an oblique circular cone, given h_1, the longest slant height, h_2, the shortest, and a, the altitude; find r, the radius of the base. $\quad \sqrt{\tfrac{1}{2}(h_1^2+h_2^2)-a^2}$. *Ans.*

300. How many square meters of canvas are required to make a conical tent which is 20 meters in diameter and 12 meters high?

Here $\quad K = \pi r h = 3{\cdot}14159 \times 10 \times \sqrt{144+100}$.

$$K = 31{\cdot}4159 \times 15{\cdot}6205$$
$$= 490{\cdot}7320+ \text{ square meters. } Ans.$$

61. $F = \tfrac{1}{2}h(p_1+p_2)$.

301. Given a basal edge b_1, and a top edge b_2, of the frustum of a regular tetrahedron; also a, the altitude of the frustum. Find h, its slant height, and F, its mantel.

$$h = \tfrac{1}{2}\sqrt{\tfrac{1}{3}(b_1-b_2)^2+4a^2}.$$
$$F = \tfrac{3}{4}(b_1+b_2)\sqrt{\tfrac{1}{3}(b_1-b_2)^2+4a^2}. \ Ans.$$

302. Same for a regular four-sided pyramid.

$$h = \tfrac{1}{2}\sqrt{(b_1-b_2)^2+4a^2}. \ Ans.$$
$$F = (b_1+b_2)\sqrt{(b_1-b_2)^2+4a^2}.$$

62. $F = \frac{1}{2}h(c_1 + c_2) = \pi h(r_1 + r_2).$

303. In the frustum of a right circular cone, given r_1, r_2, and a; find h.

304. In the frustum of a right circular cone, on each base stands a cone with its vertex in the center of the other base; from the basal radii r_1 and r_2 find the radius of the circle in which the two cones cut. $\qquad \dfrac{r_1 r_2}{r_1 + r_2}.$ *Ans.*

305. Given b_1, b_2, the basal edges, and l, the lateral edge of a frustum of a regular square pyramid; the frustum of a cone is so constructed that its upper base circumscribes the upper base of this pyramid-frustum, while its lower base is inscribed in its lower base. Find the slant height of the cone-frustum. $\qquad \sqrt{l^2 - \frac{1}{4}b_1^2 + (1 - \frac{1}{2}\sqrt{2})b_1 b_2}.$ *Ans.*

306. How far from the vertex is the cross-section which halves the mantel of a right cone? $\qquad \frac{1}{2}a\sqrt{2}.$ *Ans.*

307. Reckon the mantel from the two radii when the inclination of a slant height to one base is $45°$.
$$(r_1^2 - r_2^2)\pi\sqrt{2}. \quad Ans.$$

308. If in the frustum of a right cone the diameter of the upper base equals the slant height, reckon the mantel from the altitude a and the perimeter p of an axial section.
$$\frac{\pi}{36}(p^2 + 12a^2 + p\sqrt{p^2 - 12a^2}). \quad Ans.$$

63. $F = 2\pi aj.$

309. In the frustum of a cone of revolution, given r_1, r_2, h; find a.

310. Find the altitude of the frustum of revolution from the mantel k and the bases B_1 and B_2.
$$a = \sqrt{\frac{k - (B_1 - B_2)^2}{\pi(\sqrt{B_1} + \sqrt{B_2})}}. \quad Ans.$$

311. A right-angled triangle is revolved about an axis parallel to, and at the distance r from its side a; the areas of the circles described by its base are as m to n. Find the whole surface described by the triangle.

$$r\pi\left[r\left(\frac{m}{n}-1\right)+2a+\left(\sqrt{\frac{m}{n}}+1\right)\sqrt{a^2+r^2\left(\sqrt{\frac{m}{n}}-1\right)^2}\right]. \text{ Ans.}$$

64. $H = 4r^2\pi$.

312. Find the surface of a cube inscribed in a sphere whose surface is H.

313. A sphere is to the entire surface of its circumscribing cylinder as 2 is to 3.

314. Given r_1 and r_2, the radii of two section-circles of a sphere, and the ratio $(m:n)$ of their distances from its center. Find its radius. $\quad r = \sqrt{\dfrac{m_2 r_1^2 - n_2 r_2^2}{m^2 - n^2}}$. Ans.

315. Find the sphere whose radius is 12·6156 meters.
2000. Ans.

316. Find the sphere whose radius is $19\tfrac{125}{132}$ meters.
5000. Ans.

317. A sphere is 50·265 square meters; find its radius.
2 meters. Ans.

318. Find a sphere from a section-circle c whose distance from the center is τ. $\quad\left(\dfrac{c^2}{\pi^2}+4\tau\right)\pi$. Ans.

319. What will it cost to gild a sphere of 22·6 centimeters radius, if 100 square centimeters cost $87\tfrac{1}{2}$ cents?
$56·16. Ans.

320. Find the ratios of the mantel of the cone, described by rotating an equilateral triangle about its altitude, to the sphere generated by the circle inscribed in this triangle.
$3:2$. Ans.

65. $Z = 2\pi r a.$

321. Cut a sphere into n equal parts by parallel circles.

322. In a calot,

    (1) Given r and a; find r_1.
    (2) Given r and r_1; find Z_1.
    (3) Given a and r_1; find Z_1.
    (4) Given a and H; find Z_1. $a\sqrt{H\pi}$. *Ans.*

323. In a segment 6 centimeters high, the radii of base and top are 9 and 3 centimeters. Find area of the zone.
$$36\pi\sqrt{10} \text{ square centimeters. } Ans.$$

324. In a segment of altitude a, and congruent bases, calling the top and base radii r_1, find the zone.
$$\pi a\sqrt{4r_1^2 + a^2}. \quad Ans.$$

325. How far above the surface of the earth must a person be raised to see one-third of its surface?

Here $\quad a = \tfrac{1}{3}d = \tfrac{2}{3}r;$

and, by similar triangles,

$$x + r : r = r : r - a.$$
$$\therefore \tfrac{1}{3}r(x+r) = r^2. \quad \therefore \tfrac{1}{3}(x+r) = r.$$

$$\therefore x = 2r = d. \quad Ans.$$

326. A luminous point is distant r from a sphere of radius r; how large is the lighted surface? $\dfrac{2rr^2\pi}{r+r}$. *Ans.*

327. Find a zone from the radii of its bases r_1, r_2, and the radius of the sphere r. $\quad 2r\pi[\sqrt{r^2 - r_2^2} \pm \sqrt{r^2 - r_1^2}]$. *Ans.*

328. How far from the center must a plane be passed to divide a hemisphere into equal zone and calot?

66. Theorem of Pappus.

329. An acute-angled triangle is revolved about each side as axis; express the ratio of the surfaces of the three double-cones in terms of a, b, c, the sides of the triangle.
$$\frac{a+b}{c} : \frac{a+c}{b} : \frac{b+c}{a}. \ Ans.$$

330. The sides of a symmetric trapezoid are b_1, b_2, and z. Express the surface generated by rotating the trapezoid about one of the non-parallel sides.
$$\frac{\pi}{z}(b_1^2 + b_2^2 + b_1 z + b_2 z)\sqrt{z_2 - \tfrac{1}{4}(b_1 - b_2)^2}. \ Ans.$$

67. $O = 4\pi^2 r_1 r_2$.

331. An equilateral triangle rotates about an axis without it, parallel to, and at a distance a from one of its sides b. Find the surface thus generated. $b\pi(b\sqrt{3} + 6a). \ Ans.$

332. A rectangle with sides a and b is revolved about an axis through one of its vertices, and parallel to a diagonal. Find the generated surface. $\dfrac{4ab\pi(a+b)}{\sqrt{a^2+b^2}}. \ Ans.$

§ (L). Spherics and Solid Angles.

68. $h = 2r^2 u$.

333. Find the area of a lune whose angle is 36°.
$\tfrac{2}{5}r^2\pi. \ Ans.$

334. Find lune of 36° when $r = 1\cdot 26156$. $2. \ Ans.$

69.

335. A conical sector is one-fourth of a globe; find its solid angle. $90°. \ Ans.$
Find the vertex-angle of an axial section. $120°. \ Ans.$

HINT. By 65, Cor. 2, generating arc = 60°.

70. $\widehat{\Delta} = er^2.$

336. If two angles of a spherical triangle be right, its area varies as the third angle.

337. In a cube each solid angle is one-eighth of a steregon. (For eight cubes may be placed together, touching at a point.)

338. Find the ratio of the solid angle of a regular right triangular prism to the solid angle of a quader. $2:3$. *Ans.*

339. Find the ratio of the trihedral angles of two regular right prisms of m and n sides. $\dfrac{(m-2)n}{(n-2)m}$. *Ans.*

340. Find the area of a spherical triangle from the radius r, and the angles $\alpha = 20° 9' 30''$, $\beta = 55° 53' 32''$, $\gamma = 114° 20' 14''$. $0.1813 r^2$. *Ans.*

341. Given r, and $\alpha = 73° 12' 8''$, $\beta = 85° 3' 14''$, $\gamma = 32° 9' 16''$; find $\widehat{\Delta}$. $0.18593 r^2$. *Ans.*

342. Given r, and $\alpha = 114° 20' 5''.92$, $\beta = 30° 57' 18''.41$, $\gamma = 90° 9' 41''.67$; find $\widehat{\Delta}$. $0.9678 r^2$. *Ans.*

343. Spherical triangles on the same base are equivalent if their vertices lie in a circumference passing through the opposite extremities of sphere-diameters from the ends of the base.

344. All trihedral angles having two edges common, and their third edges *prolongations* of elements of a right cone containing the two common edges, are equivalent.

Proof: On the edges of a trihedral angle take SA, SB, SC equal; and pass through the three extremities a circle ABC of center O. Join SO, and suppose three planes to start from SO and to pass one through each edge of the trihedral angle. These planes form three new trihedrals having a common summit S, and one common edge SO. In each of these are a pair of equal dihedral angles, since each stands on an isosceles triangle with vertex at O. Thus,

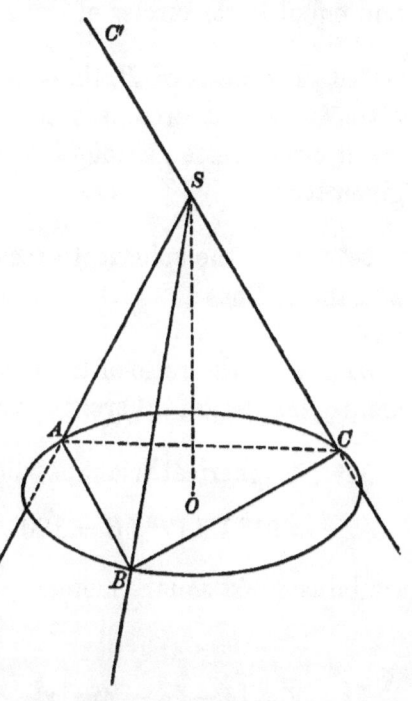

$$BA_sO = AB_sO \quad . \quad (1)$$
$$CB_sO = BC_sO \quad . \quad (2)$$
$$AC_sO = CA_sO \quad . \quad (3)$$

Therefore,

$$AB_sC + CA_sB - BC_sA$$
$$= AB_sO + CB_sO$$
$$\quad + CA_sO + BA_sO$$
$$\quad - BC_sO - AC_sO$$
$$= AB_sO + BA_sO$$
$$= 2BA_sO \; . \; . \; . \; (4)$$

Now, $2BA_sO$ remains constant as long as the summit S of the trihedral, the two edges BS and AS, and the center O of the circle are unchanged; and equation (4) holds as long as the edge SC passes through the circumference.

But
$$AB_sC = \tfrac{1}{2}\Omega - AB_sC' \; . \; . \; . \; . \; . \; . \; (5)$$
$$BA_sC = \tfrac{1}{2}\Omega - BA_sC' \; . \; . \; . \; . \; . \; . \; (6)$$
$$BC_sA = BC'_sA \; . \; . \; . \; . \; . \; . \; . \; (7)$$

Making the substitutions (5), (6), (7), equation (4) becomes

$$AB_sC' + BA_sC' + BC'_sA = \Omega - 2BA_sO.$$

The second member is constant; therefore, in the trihedral $SABC'$, the sum of the three interior angles, and consequently the area of its intercepted spherical triangle, is constant.

345. Equivalent spherical triangles upon the same base, and on the same side of it, are between the same parallel and equal lesser circles of the sphere.

346. The locus of B, the vertex of a spherical triangle of given base and area, is a lesser circle equal to a parallel lesser circle passed through A and C, the extremities of the given base.

347. Find the spherical excess of a triangle from its area and the radius. $\quad e = \dfrac{\widehat{\triangle}}{r^2 \pi} 180°$. *Ans.*

348. Find the ratio of the spherical excesses of two equivalent triangles on different spheres. $\quad e_1 : e_2 = r_2^2 : r_1^2$. *Ans.*

349. A spherical triangle whose

$$a = 91° 12' 17'', \quad \beta = 120° 9' 41'', \quad \gamma = 100° 42' 2'',$$

contains 3,962 square meters. Find the sphere.
21,600 square meters. *Ans.*

71. $\widehat{N} = [\tfrac{s}{} - (n-2)\pi] r^2$.

350. Find the ratio of the vertical solid angles of two regular pyramids of m and n sides, having the inclinations of two contiguous faces respectively, a and β.
$$\dfrac{2\pi - m(\pi - a)}{2\pi - n(\pi - \beta)}. \quad Ans.$$

351. What is the area of a spherical pentegon on a sphere of radius 5 meters, supposing the sum of the angles 640°?
43·633 square meters. *Ans.*

Exercises and Problems on Chapter V.

72. $U = abl$.

352. The diagonal of a cube is n; find its volume.

353. Find the volume of a cube whose surface is 3·9402 square meters.

354. The edge of a cube is n; approximate to the edge of a cube twice as large.

355. Find the edge of a cube equal to three whose edges are a, b, l.

356. Find the cube whose volume equals its superficial area.

357. If a cubical block of marble, of which the edge is 1 meter, costs 1 dollar, what costs a cubical block whose edge is equal to the diagonal of the first block.

$3\sqrt{3}$ dollars. *Ans.*

358. In any quader,

(1) Given a, b, and mantel; find U.

(2) Given a, b, U; find l.

(3) Given U, B, and (ab); find l and b.

(4) Given $U, \left(\dfrac{a}{b}\right), \left(\dfrac{b}{l}\right)$; find a and b.

(5) Given $(ab), (al), (bl)$; find a and b.

359. If 97 centimeters is the diagonal of a quader with square base of 43 centimeters side, find its volume.

Hint. $a^2 = (97)^2 - 2(43)^2$.

360. What weight will keep under water a cork quader 55 centimeters long, 43 centimeters broad, and 97 centimeters thick, density 0·24?

$$229\text{·}405 - 55\text{·}0572 \text{ kilograms. } Ans.$$

361. The volume of a quader whose basal edges are 12 and 4 meters is equal to the superficial area. Find its altitude.

362. In a quader of 360 square meters superficial area the base is a square of 6 meters edge. Find the volume.

363. A quader of 864 cubic centimeters volume has a square base equal to the area of two adjacent sides. Find its three dimensions.

364. In a quader of 8 meters altitude and 160 square meters surface the base is square. Find the volume.

365. The volume of a quader is 144 cubic centimeters; its diagonal 13 centimeters; the diagonal of its base 5 centimeters. Find its three dimensions.

366. In a quader of 108 square meters surface the base equals the mantel. Find volume.

367. If in the three edges of a quader, which meet in an angle, the distances of three points A, B, and C from that angle be a, b, c; then triangle $ABC = \frac{1}{2}\sqrt{a^2b^2 + a^2c^2 + b^2c^2}$.

368. How many square meters of metal will be required to construct a rectangular tank (open at top) 12 meters long, 10 meters broad, and 8 meters deep. 472. *Ans.*

369. The three external edges of a box are 3, 2·52, and 1·523 meters. It is constructed of a material 0·1 meters in thickness. Find the cubic space inside.

$$8\text{·}594208 \text{ cubic meters. } Ans.$$

73. $\delta = \dfrac{\omega^g}{V^{ccm}} = \dfrac{\omega^{kg}}{V^l}.$

370. A brick 11 centimeters long, 3 centimeters broad, 2 centimeters thick, weighs 45 grams; find its density.

371. A cube of pine wood of 12 centimeters edge weighs 1 kilogram; find the density of pine. 0·57. *Ans.*

372. If a mass of ice containing 270 cubic meters weighs 229,000 kilograms, find the density of ice. 0·92. *Ans.*

373. If a cubic centimeter of metal weighs 6·9 grams, what is its density?

74. V. $P = abl.$

374. If the base of a parallelepiped is a square, find the altitude a and basal edge b from the volume and mantel.

75. V. $P = aB.$

375. The base of a prism 10 meters tall is an isosceles right triangle of 6 meters hypothenuse; find volume.

376. In a prism whose base is 210 square meters, the three sides are rectangles of 336, 300, 204 square meters; find volume.

377. Find altitude of a right prism of 480 cubic centimeters volume, standing upon an isosceles triangle whose base is 10 centimeters and side 13 centimeters.

378. In a right prism of 54 cubic centimeters volume, the mantel is four times the base, an equilateral triangle; find basal edge.

379. The vertical ends of a hollow trough are parallel equilateral triangles, with 1 meter in each side, the bases of the triangles being horizontal. If the distance between the triangular ends be 6 meters, find the number of cubic meters of water the trough will contain.

2·598 cubic meters. *Ans.*

76. V. C $= ar^2\pi$.

380. In a right circular cylinder,

(1) Given a and c;   find V. C. $\dfrac{ac^2}{4\pi}$. *Ans.*
(2) Given a and C;   find V. C.
(3) Given (V. C) and C; find r.

$$V.\,C = ar^2\pi, \quad \therefore a = \frac{V.\,C}{r^2\pi}.$$

$$C = a2r\pi, \quad \therefore a = \frac{C}{2r\pi}.$$

$$\therefore \frac{V.\,C}{r^2\pi} = \frac{C}{2r\pi}. \quad \therefore r = \frac{2V.\,C}{C}. \quad Ans.$$

381. If $C = 91\cdot84$ square meters, and V. C $= 145$ cubic meters, find a. $\qquad a = 4\cdot628986$ meters. *Ans.*

382. A right cylinder of 50 cubic centimeters volume has a circumference of 9 centimeters; find mantel and volume.

383. In a right cylinder of 8 cubic centimeters the mantel equals the sum of the bases; find altitude.

384. If, in three cylinders of the same height, one radius is the sum of the other two, then one curved surface is the sum of the others, but contains a greater volume.

385. Find the ratio of two cylinders when the radius of one equals the altitude of the other.

386. Find the ratio of two cylinders whose mantels are equivalent.

387. If 1728 cubic meters of brass were to be drawn into wire of one-thirtieth of a meter in diameter, determine the length of the wire.

Here
$$1728 = ar^2\pi = a\left(\frac{1}{60}\right)^2\pi = \frac{a\pi}{3600}.$$
$$\therefore a = \frac{1728 \times 3600}{\pi} = 1{,}980{,}145 \text{ meters. } Ans.$$

388. What must be the ratio of the radius of a right cylinder to its altitude, in order that the axis-section may equal the base? $2:\pi$. *Ans.*

389. A cylindric glass of 5 meters diameter holds half a liter; find its height.

390. A rectangle whose sides are 3 meters and 6 meters is turned about the 6-meter side as axis; find the volume of the generated cylinder.

391. The diagonal of the axis-section of a right cylinder is 5 centimeters; the diameter of its base is three-fourths its height. Find its volume.

392. In a right cylinder, from A, the area of the axis-section, reckon the area of that section which halves the basal radius normal to it. $\tfrac{1}{2}A\sqrt{3}$. *Ans.*

393. The longest side of a truncated circular cylinder of 1·5 meters radius is 2 meters; the shortest, 1·75 meters. Find volume.

394. If a room be 40 meters long by 20 meters broad, what addition will be made to its cubic contents by throwing out a semicircular bow at one end?
2513.28 cubic meters. *Ans.*

395. The French and German liquid measures must be cylinders of altitude twice diameter. Find the altitude for measures holding 2 liters, 1 liter, and ½ liter.
216·7, 172·1, and 136·5 millimeters. *Ans.*

396. The German dry measures must be cylinders of altitude two-thirds diameter. Find diameter of a measure containing 100 liters. 575·9 millimeters. *Ans.*

397. In the French grain measure the altitude equals diameter. Find for hectoliter. 503·7 millimeters. *Ans.*

77. $V. C_1 - V. C_2 = a\pi (r_1 + r_2)(r_1 - r_2)$.

398. How many cubic meters of iron are there in a roller which is half a meter thick, with an outer circumference of 61 meters, and a width of 37 meters? ($\pi = \tfrac{22}{7}$).
1353 cubic meters. *Ans.*

399. Find the amount of metal in a pipe 3·1831 meters long, with $r_1 = 12$ meters and $r_2 = 8$ meters.
800 cubic meters. *Ans.*

400. The amount of metal in a pipe is 175·9292 cubic meters, its length is 3·5 meters, and its greater radius is 5 meters. Find its thickness. 2 meters. *Ans.*

78.   Sections Similar.

401. A regular square pyramid, whose basal edge is b, is so cut parallel to the base that the altitude is halved; find the area of this cross-section.

402. A section parallel to the base of a cone (base-radius r), its altitude in the ratio of m to n. Find the area of this section. $\dfrac{m^2 r^2 \pi}{(m+n)^2}$. *Ans.*

403. On each of the bases of a right cylinder, radius r, stands a cone whose vertex is the center of the other base. Find the circumference in which the cone-mantels cut.
$r\pi$. *Ans.*

79. Equivalent Tetrahedra.

404. If a plane be drawn through the points of bisection of two opposite edges of a tetrahedron, it will bisect the tetrahedron.

80. V. Y $= \tfrac{1}{3} aB$.

405. A pyramid of 9 decimeters altitude contains $15\tfrac{3}{4}$ cubic meters; find its base. 52.5 square meters. *Ans.*

406. The pyramid of Memphis has an altitude of 73 Toises; the base is a square whose side is 116 Toises. If a Toise is 1·95 meters, find the volume of this pyramid.
About 2,427,780 cubic meters. *Ans.*

407. A goldsmith uses up a triangular pyramid of gold, density 19·325, and charges $900 a kilogram. What is his bill if the altitude of the pyramid is 4 centimeters, the altitude of its base 4 millimeters, and the base of its base 1·5 centimeters. $6·97$\tfrac{1}{2}$. *Ans.*

408. Find the volume of a pyramid of 30 meters altitude, having for base a right triangle of 25 meters hypothenuse and 7 meters altitude.

81. V. K $= \tfrac{1}{3} ar^2 \pi$.

409. In a right circular cone,

(1) Given r and h; find V. K. $\tfrac{1}{3} r^2 \pi \sqrt{h^2 - r^2}$. *Ans.*

(2) Given a and h; find V. K. $\tfrac{1}{3} a\pi (h^2 - a^2)$. *Ans.*

(3) Given r and K; find V. K. $\tfrac{1}{3} r \sqrt{K^2 - r^4 \pi^2}$. *Ans.*

(4) Given h and K; find V. K. $\dfrac{K^2}{3 h^2 \pi} \sqrt{h^2 - \dfrac{K^2}{h^2 \pi^2}}$. *Ans.*

(5) Given a and K; find V. K. $\tfrac{1}{3} \pi a \left[\sqrt{\dfrac{k^2}{\pi^2} + \dfrac{a^4}{4} - \dfrac{a^2}{2}} \right]$. *Ans.*

410. A cone and cylinder have equal surfaces, and their axis-sections are equilateral; find the ratio of their volumes.

$$\text{Surface of cylinder} = \frac{d^2\pi}{2} + d^2\pi = \frac{3d^2\pi}{2}.$$
$$\text{Surface of cone} \quad = \frac{h^2\pi}{4} + \frac{h^2\pi}{2} = \frac{3h^2\pi}{4}.$$
$$\therefore h = d\sqrt{2}.$$

411. In a triangular prism of 9 meters altitude, whose base has 4 square meters area and 8·85437 meters perimeter, a cylinder is inscribed. Find the base and altitude of an equivalent cone whose axial section is equilateral.
$B = 17\cdot1236$ square meters, $a = 4\cdot043738$ meters. *Ans.*

412. Find the edge of an equilateral cone holding a liter.
16·4 centimeters. *Ans.*

413. Halve an equilateral cone by a plane parallel to the base.

414. Find the ratio of the volumes of the cones inscribed and circumscribed to a regular tetrahedron whose edge is n.

$$\frac{\left(\frac{n}{2}\sqrt{\frac{1}{3}}\right)^2}{\left(\frac{n}{\sqrt{3}}\right)^2}. \quad Ans.$$

82. Prismoidal Formula: $D = \frac{1}{6}a(B_1 + 4M + B_2)$.

415. Find the volume of a rectangular prismoid of 12 meters altitude, whose top is 5 meters long and 2 meters broad, and base 7 meters long and 4 meters broad.
220 cubic meters. *Ans.*

416. In a prismoid 15 meters tall, whose base is 36 square meters, the basal edge is to the top as 3 to 2. Find the volume.
380 cubic meters. *Ans.*

EXERCISES AND PROBLEMS.

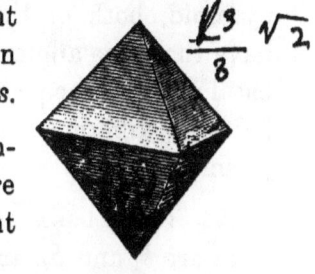

417. Every regular octahedron is a prismatoid whose bases and lateral faces are all congruent equilateral triangles. Find its volume in terms of an edge b. $\quad$ *Ans.* $\dfrac{b^3}{3}\sqrt{2}$.

418. The bases of a prismatoid are congruent squares of side b, whose sides are not parallel; the lateral faces are eight isosceles triangles. Find the volume.
$\frac{1}{3}ab^2(2+\sqrt{2})$. *Ans.*

419. If, from a regular icosahedron, we take off two five-sided pyramids whose vertices are opposite summits, there remains a solid bounded by two congruent regular pentagons and ten equilateral triangles. Find its volume from an edge b.
$\frac{1}{6}b^3(5+2\sqrt{5}.)$ *Ans.*

420. Both bases of a prismatoid of altitude a are squares; the lateral faces isosceles triangles; the sides of the upper base are parallel to the diagonals of the lower base, and half as long as these diagonals; b is a side of the lower base. Find the volume. $\quad\frac{5}{6}ab^2$. *Ans.*

421. The upper base of a prismatoid of altitude $a = 6$ is a square of side $b_2 = 7{\cdot}07107$; the lower base is a square of side $b_1 = 10$, with its diagonals parallel to sides of the upper base; the lateral faces are isosceles triangles. Find volume.
$\frac{1}{3}a(b_1^2 + b_1 b_2\sqrt{2} + b_2^2) = 500$. *Ans.*

422. Every prismatoid is equivalent to three pyramids of the same altitude with it, of which one has for base half the sum of the prismatoid's bases, and each of the others its midcross section.
$$D = \tfrac{1}{3}a\left(\frac{B_1 + B_2}{2} + 2M\right).$$

210   MENSURATION.

423. Every prismoid is equivalent to a prism plus a pyramid, both of the same altitude with it, whose bases have the same angles as the bases of the prismoid; but the basal edges of the prism are half the sum, and of the pyramid half the difference, of the corresponding sides of both the prismoid's bases.

424. If the bases of a prismoid are trapezoids whose midlines are b_1 and b_2, and whose altitudes are a_1 and a_2,

$$D = a\left(\frac{a_1+a_2}{2} \cdot \frac{b_1+b_2}{2} + \tfrac{1}{3} \cdot \frac{a_1-a_2}{2} \cdot \frac{b_1-b_2}{2}\right).$$

83. V. $F = \tfrac{1}{3}a(B_1 + \sqrt{B_1 B_2} + B_2)$.

425. A side of the base of a frustum of a square pyramid is 25 meters, a side of the top is 9 meters, and the height is 240 meters. Required the volume of the frustum.
Here   V. $F = \tfrac{1}{3} 240 (625 + 225 + 81)$
              $= 80 \times 931 = 74{,}480$ cubic meters. *Ans.*

426. The sides of the square bases of a frustum are 50 and 40 centimeters; each lateral edge is 30 centimeters. Find the volume.   59·28 liters. *Ans.*

427. In the frustum of a pyramid whose base is 50 square meters, and altitude 6 meters, the basal edge is to the corresponding top edge as 5 to 3. Find volume.
                                       588 cubic meters. *Ans.*

428. Near Memphis stands a frustum whose height is 142·85 meters, and bases are squares on edges of 185·5 and 3·714 meters. Find its volume.

429. In the frustum of a regular pyramid, volume is 327 cubic meters, altitude 9 meters, and sum of basal and top edge 12 meters. Find these. 7 meters and 5 meters. *Ans.*

EXERCISES AND PROBLEMS. 211

430. In the frustrum of a regular tetrahedron, given a basal edge, a top edge, and the volume. Find the altitude.

84. V. $F = \frac{1}{3}a\pi(r_1^2 + r_1 r_2 + r_2^2)$.

431. Divide a cone whose altitude is 20 into three equivalent parts by planes parallel to the base.

Volume of whole cone $= \frac{1}{3}r^2\pi\, 20$.
Volume of midcone $= \frac{2}{3}r_1^2\pi(20-a)$.

$$\therefore r_1^2(20-a) = \frac{2}{3}r^2\, 20.$$

But $\qquad r : 20 = r_1 : 20-a$.

$$\therefore r_1 = \frac{20-a}{20}r.$$

$$\therefore \frac{(20-a)^3}{400} = \frac{40}{3}.$$

$$\therefore (20-a)^3 = \frac{16000}{3} = 5333\cdot333+.$$

$$\therefore 20-a = \sqrt[3]{5333\cdot333+} = 17\cdot471+.$$

$$\therefore a = 2\cdot528+.$$

In the same way, $\qquad a' = 3\cdot604+$.

THEOREM OF CLAVIUS.

432. The frustum of a cone equals the sum of a cylinder and cone of frustral altitude whose radii are respectively the half-sum and half-difference of the frustral radii.

$$V.\ F = a\pi\left(\frac{r_1+r_2}{2}\right)^2 + \frac{1}{3}a\pi\left(\frac{r_1-r_2}{2}\right)^2.$$

This is a formula convenient for computation.

433. A frustum of 8 meters altitude, with $r_1 = 4$ and $r_2 = 2$, is halved by a plane parallel to the base. Find radius of section and its distance from top of frustum.

$$r_3 = \sqrt{36};\ \ a_3 = 4\sqrt{36} - 8.\ \ Ans.$$

434. In a frustum of 3 meters altitude and 63 cubic meters volume, $r_1 = 2r_2$; find r_2.
$$3\sqrt{\frac{1}{\pi}} \quad Ans.$$

435. In the frustum where $a = 8$ meters, $r_1 = 10$ meters, $r_2 = 6$ meters, the altitude is cut into four equal parts by planes parallel to the base. Find the radii of these sections.

HINT. The altitude of the completed cone is $8 + 12 = 20$, and of the others, 18, 16, 14. ∴ by similar triangles,
$$20:18:16:14::10:9:8:7.$$
7, 8, 9 meters. *Ans.*

436. The frustum of an equilateral cone contains 2 hectoliters, and is 40 centimeters in altitude. Find the radii.
27·785 and 50·879 centimeters. *Ans.*

85. PRISMOIDAL FORMULA: $D = \frac{1}{6} a (B_1 + 4M + B_2)$.

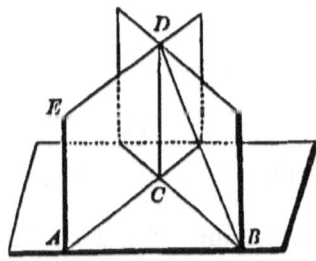

437. A solid is bounded by the triangles ABC, CBD, the parallelogram $ACDE$, and the skew quadrilateral $BAED$ whose elements are parallel to the plane BCD. Find its volume. $\frac{1}{2} a . ABC$. *Ans.*

The skew quadrilateral is part of a warped surface called the hyperbolic paraboloid.

438. A tetrahedron is bisected by the hyperbolic paraboloid whose directrices are two opposite edges, and whose plane director is parallel to another pair of opposite edges.

439. A solid is bounded by a parallelogram, two skew quadrilaterals, and two parallel triangles; find its volume.
$\frac{1}{3} a (\triangle_1 + \triangle_2)$. *Ans.*

EXERCISES AND PROBLEMS. 213

440. Twice the volume of the segment of a ruled surface between parallel planes is equivalent to the sum of the cylinders on its bases, diminished by the cone whose vertex is in one of the parallel planes, and whose elements are respectively parallel to the lines of the ruled surface.

86. $W = \frac{1}{6} aw(2b_1 + b_2)$.

441. A wedge of 10 centimeters altitude, 4 centimeters edge, has a square base of 36 centimeters perimeter. Find volume.

442. The three parallel sides of a truncated prism are 8, 9, and 11 meters. The section at right angles to them is a right-angled triangle, with hypothenuse 17 meters, and one side 15 meters. Find volume.

443. The volume of a truncated regular prism is equal to the area of a right section multiplied by the axis or *mean length* of all the lateral edges.

$$V = A \frac{l_1 + l_2 + \cdots + l_n}{n}.$$

444. To find the volume of any truncated prism.

Rule: Multiply the length of each edge by the sum of the areas of all the triangles in the right section which have an angular point in that edge. The sum of the products will be three times the volume.

Formula: $3V = \Sigma \Delta l$.

87. $X = \frac{2}{3} aM$.

445. Given V, the volume of a parallelepiped; in each of two parallel faces draw a diagonal, so that the two diagonals cross. Take the ends of these as summits of a tetrahedron, and find its volume. $\frac{1}{3} V$. *Ans.*

88. V. H $= \frac{4}{3}\pi r^3$.

446. Find the volume of a sphere whose superficial area is 20 meters.

447. Find the radius of a sphere equal to the sum of two spheres whose radii are 3 and 6 centimeters.

$3\sqrt[3]{9}$ centimeters. *Ans.*

448. Find the radius of a golden globe, density 19·35, weighing a kilogram.

449. A solid metal globe 6 meters in diameter is formed into a tube 10 meters in external diameter and 4 meters in length. Find the thickness of the tube. 1 meter. *Ans.*

450. If a cone and hemisphere of equal bases and altitudes be placed with their axes parallel, and the vertex of the cone in the plane of the base of the hemisphere, and be cut by a plane normal to their axes, the sum of the sections will be a constant.

Archimedes' Theorem.

451. Cone, hemisphere, and cylinder, of same base and altitude, are as $1:2:3$.

452. The surfaces and the volumes of a sphere, a circumscribed right cylinder, and a circumscribed right cone whose axial section is an equilateral triangle, are as $4:6:9$. Therefore, the cylinder is a geometric mean between the sphere and cone.

Hint.
$H = 4r^2\pi.$ $\quad$ V. H $= \frac{4}{3}r^3\pi.$
$C + 2B = 6r^2\pi.$ $\quad$ V. C $= \frac{8}{3}r^3\pi.$
$K + B = 9r^2\pi.$ $\quad$ V. K $= \frac{9}{3}r^3\pi.$

453. A quader, having a square base of 5 centimeters edge, is partly filled with water. Into it is put an iron ball going fully under the water, which rises 1·33972 centimeters. Find the diameter of the sphere.

$$\frac{d^3\pi}{6} = 5^2 \times 1\cdot 33972.$$

89. V. G $= \tfrac{1}{6} a\pi[3(r_1^2 + r_2^2) + a^2].$

454. If a heavy globe, whose diameter is 4 meters, be let fall into a conical glass, full of water, whose diameter is 5 meters and altitude 6 meters, it is required to determine how much water will run over.

The slant height of the cone

$$h = \sqrt{36 + 6\cdot 25} = \sqrt{42\cdot 25} = 6\cdot 5.$$

If a is the altitude of the dry calot,

$$6\cdot 5 : 2\cdot 5 = 6 - (2-a) : 2.$$
$$13 = 10 + 2\cdot 5a.$$
$$\therefore a = 1\cdot 2.$$

But dry segment equals

$a^2\pi(r - \tfrac{1}{3}a) = 1\cdot 44\,\pi(2 - 0\cdot 4) = 1\cdot 44\,\pi\, 1\cdot 6 = 2\cdot 304\,\pi = 7\cdot 2382233+.$

But V. H $= \tfrac{1}{6} d^3\pi = 33\cdot 5104.$

∴ volume of segment immersed is

26·272+ cubic meters. *Ans.*

455. A section parallel to the base of a hemisphere bisects its altitude; find the ratio of the parts of the hemisphere. 5 : 11. *Ans.*

456. A sphere is divided by a plane in the ratio 5 : 7. In what ratio is the globe cut?   325 : 539. *Ans.*

457. A calot 8 centimeters high contains 1200 cubic centimeters; find radius of the sphere.

$$\frac{75}{\pi}$$ cubic centimeters. *Ans.*

458. Find the volume of a segment of 12 centimeters altitude, the radius of whose single base is 24 centimeters.
$$r = 30 \text{ centimeters};$$
$$\text{V. G} = \cdot014976\,\pi \text{ cubic meters. } Ans.$$

459. In terms of sphere-radius, find the altitude of a calot n times as large as its base. $\quad a = \left(\dfrac{n-1}{n}\right)2r.\ Ans.$

460. Find the ratio of the volume of a sphere to the volume of its segment whose calot is n times its base.
$$\text{V. H : to V. G} :: n^3 : (n-1)^2(n+2).\ Ans.$$

461. Find volume of a segment whose calot is 15·085 square meters, and base 2 meters, from sphere-center.
$$\text{V. G} = 5\cdot737025 \text{ cubic meters. } Ans.$$

462. In a sphere of 10 centimeters radius, find the radii r_1 and r_2 of the base and top of a segment whose altitude is 6 centimeters, and base 2 centimeters, from the sphere-center. $\quad r_1 = 4\sqrt{6}$ centimeters, $r_2 = 6$ centimeters. $Ans.$

463. Out of a globe of 12 centimeters radius is cut a segment whose volume is one-third the globe, and whose bases are congruent and 8 centimeters apart; find the radius of bases. $\quad r_1 = r_2 = 4\sqrt{2\left(\dfrac{3}{\pi} - \dfrac{1}{3}\right)}$ centimeters. $Ans.$

90. V. S $= \frac{2}{3}\pi a r^2$.

464. In a spherical sector,

(1) Given r, r_1, r_2; find V. S.
(2) Given a, r_1, r_2; find V. S.

465. In a sphere of radius r, find the altitude of a segment which is to its sector as n to m.
$$a = r\left(\dfrac{3}{2} \pm \sqrt{\dfrac{9}{4} - \dfrac{2n}{m}}\right).\ Ans.$$

466. A sector is $\frac{1}{n}$ of its globe, whose diameter is d; find the volume of its segment. V. $G = \frac{\pi d^3}{6}\left(\frac{3n-2}{n^3}\right)$. *Ans.*

467. A sphere of given volume V is cut into two segments whose altitudes are as m to n; find both calots z_1 and z_2, and the segments.

$$z_1 = \frac{m}{m+n}\sqrt[3]{36\pi V^2}; \quad z_2 = \frac{n}{m+n}\sqrt[3]{36\pi V^2};$$

$$G_1 = \frac{m^2(m+3n)}{(m+n)^3}V; \quad G_2 = \frac{n^2(3m+n)}{(m+n)^3}V. \quad Ans.$$

91. $\hat{v} = \frac{2}{3}r^3 u$.

468. Find the volume of a spherical ungula whose radius is 7·6 and ⚥ 18° 12′. 92·958. *Ans.*

469. ⚥ $= 26° 6′$, $r = 13·2$. Find $\hat{v}$. 698·45. *Ans.*

470. A lune of 192 square meters has radius 15 meters; find volume of the ungula. 960 cubic meters. *Ans.*

471. Given L and r; find $\hat{v}$. $\hat{v} = \frac{Lr}{3}$. *Ans.*

472. Given $\hat{v}$ and r; find ⚥. ⚥ $= \frac{270\hat{v}}{r^3\pi}$. *Ans.*

473. Given L and ⚥; find $\hat{v}$. $\hat{v} = L\sqrt{\frac{10L}{\pi \text{⚥}}}$. *Ans.*

92. $\widehat{Y} = \frac{1}{3}r^3 e$.

474. In a spherical pyramid given the angles of its triangular base, $a = 78° 15′$, $\beta = 144° 30′$, $\gamma = 108° 15′$, and given $r = 10·8$; find $\widehat{Y}$. 1106·61. *Ans.*

475. Given a, β, γ, and $\widehat{\Delta}$; find $\widehat{Y}$. $2\widehat{\Delta}\sqrt{\frac{5\widehat{\Delta}}{c\pi}}$. *Ans.*

476. $\widehat{\Delta} = 486$, $a = 84°\,13'$, $\beta = 96°\,27'$, $\gamma = 112°\,20'$; find Y. $2543\cdot06$. *Ans.*

477. Given $r = 8\cdot8$, $a = 106°\,30'$, $\beta = 120°\,10'$, $\gamma = 150°\,15'$, $\delta = 112°\,5'$; find the four-faced $\widehat{Y}$. $511\cdot433$. *Ans.*

93. Theorem of Pappus.

478. If an equilateral triangle whose sides are halved by a straight line rotates about its base, the two volumes generated are equivalent.

479. A trapezoid rotates first about the longer, then about the shorter, of its parallel sides; the volumes of the solids generated are as m to n. Find the ratio of the parallel sides. $\dfrac{2n-m}{2m-n}$. *Ans.*

94. V. $O = 2\pi^2 abr$.

480. Find the volume of a solid generated by rotating a parallelogram about an axis exterior to it; given the area of the parallelogram $\square$, and the distance r of the intersection-point of its diagonals from the axis. $2\pi r\square$. *Ans.*

481. The volume of a spiral spring, whose cross-section is a circle, equals the product of this generating circle by the length of the helix along which its center moves. The helix is the curve traced upon the surface of a circular cylinder by a point, the direction of whose motion makes a constant angle with the generating line of the cylinder.

482. A regular hexagon rotates about l, one of its sides; find the volume generated. $\tfrac{9}{2}l^3\pi$. *Ans.*

95. $V_1 = \dfrac{V_2 a_1^3}{a_2^3}.$

483. Any two similar solids may be so placed that all the lines joining pairs of homologous points intersect in a point. Every two homologous lines or surfaces in the two solids are then parallel.

484. Any two symmetric solids may be so placed that all the lines joining pairs of homologous points intersect in a point. This point bisects each sect. Every two homologous lines are then parallel.

485. Three persons having bought a sugar-loaf, would divide it equally among them by sections parallel to the base. It is required to find the altitude of each person's share, supposing the loaf to be a cone whose height is 20.
$$13 \cdot 8672, \ 3 \cdot 6044, \text{ and } 2 \cdot 5284. \ Ans.$$
Let altitude of upper cone equal x, and its volume equal 1. Now,
$$1 : 3 = x^3 : 20^3.$$
$$\therefore x = \sqrt[3]{2666 \cdot 666} = 13 \cdot 867+.$$

96.   IRREGULAR SOLIDS.

486. When a solid is placed in a square quader of basal edge 6 meters, the liquid, rising 3·97 meters, covers it; find its volume.

97. $V^{ccm} = \dfrac{\omega^g}{\delta}.$

487. How much mercury, density 13·60, will weigh 7·59 grams?

488. If the density of zinc is 7·19, find how much weighs 3·83 kilograms.

98. $I = \tfrac{1}{3}[x_2(B_1 - B_3) + x_3(B_2 - B_4) + \text{etc.}$
$+ x_n(B_{n-1} - B_{n+1}) + x_{n+1}(B_n + B_{n+1})]$
$+ \tfrac{2}{3}[x_2 M_1 + (x_3 - x_2) M_2 + (x_4 - x_3) M_3 + \text{etc.}$
$+ (x_{n+1} - x_n) M_n]$.

489. If the areas of six parallel planes 2 meters apart are 1, 3, 5, 7, 9, 11 square meters, and of the five mid-sections 2, 4, 6, 8, 10 square meters, find the whole volume.

99. $A_z = q + mx + nx^2 + fx^3$.

490. Find an expression for the volume of a semicubic paraboloid generated by the revolution of a semicubic parabola round its axis. In this curve $y^2 \propto x^3$, the revolving ordinate being y.

491. A paraboloid and a semicubic paraboloid have a common base and vertex; show that their volumes are as $2:1$.

492. A vessel, whose interior surface has the form of a prolate spheroid, is placed with its axis vertical, and filled with a fluid to a depth h; find the depth of the fluid when the axis is horizontal.

493. A square-threaded screw, with double thread, is formed upon a solid cylinder 3 meters in diameter; the thread projects from the cylinder $\tfrac{5}{16}$ of meter, and the screw rises 3 meters in four turns. Find the volume, if the screw be 9 meters in length.

494. Find the volume of a square groin, the base of which is 15 meters square, and the guiding curve a semicircle.

495. The prismoidal formula applies to any shape contained by two parallel bases, and a lateral surface generated by the motion of a parabola or cubic parabola whose plane is always parallel to a given plane, but whose curvature may pass through any series of changes in amount, direction, and position.

496. No equation of finite degree, *representing a bounding surface*, can define the limits of applicability of the prismoidal formula, because surfaces of higher degrees enclose prismoidal spaces.

100. $V = \frac{a}{4}(B_1 + 3A_{\frac{1}{3}a}) = \frac{a}{4}(B_2 + 3A_{\frac{a}{3}}).$

497. Show how existing rules for the estimation of railroad excavation may be improved.

101. $\xi = \frac{3}{10} h [5(y_2 + y_4 + y_6) + y_4 + y_1 + y_3 + y_5 + y_7].$

498. If a parabolic spindle is equal in volume to one-fifth of the sphere on its axis as diameter, show that its greatest diameter is equal to half its length.

499. A parabolic spindle is placed in a cylinder half-full of water, the greatest diameter of the spindle being equal to that of the interior of the cylinder; find the height of the cylinder so that the water may just rise to the top.

500. A vessel, laden with a cargo, floats at rest in still water, and the line of flotation is marked. Upon the removal of the cargo every part of the vessel rises 3 meters, when the line of flotation is again marked. From the known lines of the vessel the areas of the two planes of flotation and of five intermediate equidistant sections are calculated and found to be as follows, the areas being expressed in square meters: 3918, 3794, 3661, 3517, 3361, 3191, 3004. Find the weight of the cargo removed.

Exercises and Problems on Chapter VIII.

501. A square on the line b is divided into four equal triangles by its diagonals which intersect in C; if one triangle be removed, find the $^\mu C$ of the figure formed by the three remaining triangles. $\quad CL = \dfrac{b}{9}.$ Ans.

HINT. For such problems let L be the $^\mu C$ of the part left, and O of the part cut out; then

$$CL \times \text{area left} = CO \times \text{area cut out}.$$

502. If a heavy triangular slab be supported at its angles, the pressure on each prop will be one-third the weight of the slab.

503. A weight ω is placed at any point O upon a triangular table ABC (supposed without weight). Show that the pressures on the three props (viz., A, B, C) are proportional to the areas of the triangles BOC, AOC, AOB respectively.

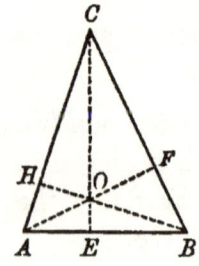

Draw the straight lines AOF, BOH, COE; and let A', B', C' be the pressures at A, B, C respectively. Then

$$C' \times CE = \omega \times OE.$$

$$\therefore \frac{C'}{\omega} = \frac{AOB}{ABC}.$$

Similarly, for A' and B'.

504. The mid-point of one side of a square is joined with the mid-points of the adjacent sides, and the triangles thus formed are cut off; find the $^\mu C$ of the remainder.

505. If two triangles stand on the same base, the line joining their μC's is parallel to the line joining their vertices.

506. Find the distance from the base of the μC of four uniform rods forming a trapezoid, the two parallel sides of which are respectively 12 meters and 30 meters long, and the other sides each 15 meters long. $5\frac{1}{4}$ meters. *Ans.*

507. The altitude of the segment of a globe is a; find height of μC of its zone. $\frac{1}{2}a$. *Ans.*

508. Find μC of a hemisphere.

509. Find μC of cylinder-mantel.

510. Find μC of cone-mantel.

511. If a body of density δ weighs ω, express the distance of its μC from its midcross-section. $\dfrac{a^2(B_2 - B_1)\delta}{12\omega}$. *Ans.*

512. Find the μC of a portion of a parabola cut off by a line perpendicular to the axis at a distance h from the vertex. $\frac{3}{5}h$. *Ans.*

513. Find the μC of the segment of a globe at a distance b from the center. $\dfrac{3(r+b)^2}{4(2r+b)}$. *Ans.*

514. Find the distance from vertex of the μC of half a prolate spheroid.

515. A right circular cone, whose vertical angle is 60°, is constructed on the base of a hemisphere; find the μC of the whole body.

516. Show that the compound body of the last exercise will rest in any position on its convex spherical surface.

517. Every body or system of particles has a μC, and cannot have more than one.

518. Find the $^\mu C$ of any polygon by dividing it into triangles.

519. If the sides of a triangle be 3, 4, and 5 meters, find the distance of $^\mu C$ from each side. $\frac{4}{3}$, 1, $\frac{2}{3}$ meter. *Ans.*

Miscellaneous.

520. Find both sides of a rectangle from their ratio $m:n$, and its area R.
$$a = \sqrt{\frac{mR}{n}}; \; b = \sqrt{\frac{nR}{m}}. \; Ans.$$

521. If two triangles have one angle of the one equal to one angle of the other, and the sides about a second angle in each equal, then the third angles will be either equal or supplemental.

522. Two triangles are congruent, if two sides and a medial in the one are respectively equal to two sides and a corresponding medial in the other.

523. Two triangles are congruent, if three medials in one equal those in the other.

524. On a plane lie three tangent spheres of radius r; upon these lies a fourth of radius r'. How high is its center above the plane, and how large at least is r', since the sphere does not fall through?

LOGARITHMS.

133. The logarithm a of a number n to a given base b is the index of the power to which the base must be raised to give the number:

So, if $b^a = n$, then $^b\log n = a$, or the b-logarithm of n is a.

134. $\quad\quad\quad ^b\log b = 1. \quad ^b\log 1 = 0.$

135. $\quad\quad\quad ^b\log mn = {^b\log m} + {^b\log n}.$

136. $\quad\quad\quad ^b\log \dfrac{m}{n} = {^b\log m} - {^b\log n}.$

137. $\quad\quad\quad ^b\log n^p = p \times {^b\log n}.$

138. $\quad\quad\quad ^b\log n^{\frac{1}{p}} = \dfrac{1}{p} \times {^b\log n}.$

139. $\quad\quad\quad ^{b'}\log n = {^b\log n} \times \dfrac{1}{^b\log b'}.$

$\dfrac{1}{^b\log b'}$ is called the *modulus* or multiplier for transforming the log of a number to base b to the log of same number to base b'.

140. The base of the common system of logarithms is 10.

$$^{10}\log(n \times 10^p) = {^{10}\log n} + p.$$

141. $\quad\quad\quad ^{10}\log(n \div 10^p) = {^{10}\log n} - p.$

142. The *mantissa* is the decimal part of a logarithm. The *characteristic* is the integral part of a logarithm.

The logs of all numbers consisting of the same digits in the same order have the same mantissa.

143. The characteristic of the log of a number is *one less* than the number of digits in the integral part.

144. When the number has no integral figures, the characteristic of its log is negative, and is *one more* than the number of cyphers which precede the first significant digit; that is, the number of cyphers (zeros) immediately after the decimal point.

LOGARITHMS. 227

N	0	1	2	3	4	5	6	7	8	9	PP
10	0000	0043	0086	0128	0170	0212	0253	0294	0334	0374	4.8
11	0414	0453	0492	0531	0569	0607	0645	0682	0719	9755	4.8
12	0792	0828	0864	0899	0934	0969	1004	1038	1072	1106	3.7
13	1139	1173	1206	1239	1271	1303	1335	1367	1399	1430	3.6
14	1461	1492	1523	1553	1584	1614	1644	1673	1703	1732	3.6
15	1761	1790	1818	1847	1875	1903	1931	1959	1987	2014	3.6
16	2041	2068	2095	2122	2148	2175	2201	2227	2253	2279	3.5
17	2304	2330	2355	2380	2405	2430	2455	2480	2504	2529	2.5
18	2553	2577	2601	2625	2648	2672	2695	2718	2742	2765	2.5
19	2788	2810	2833	2856	2878	2900	2923	2945	2967	2989	2.4
20	3010	3032	3054	3075	3096	3118	3139	3160	3181	3201	2.4
21	3222	3243	3263	3284	3304	3324	3345	3365	3385	3404	2.4
22	3424	3444	3464	3483	3502	3522	3541	3560	3579	3598	2.4
23	3617	3636	3655	3674	3692	3711	3729	3747	3766	3784	2.4
24	3802	3820	3838	3856	3874	3892	3909	3927	3945	3962	2.4
25	3979	3997	4014	4031	4048	4065	4082	4099	4116	4133	2.3
26	4150	4166	4183	4200	4216	4232	4249	4265	4281	4298	2.3
27	4314	4330	4346	4362	4378	4393	4409	4425	4440	4456	2.3
28	4472	4487	4502	4518	4533	4548	4564	4579	4594	4609	2.3
29	4624	4639	4654	4669	4683	4698	4713	4728	4742	4757	1.3
30	4771	4786	4800	4814	4829	4843	4857	4871	4886	4900	1.3
31	4914	4928	4942	4955	4969	4983	4997	5011	5024	5038	1.3
32	5051	5065	5079	5092	5105	5119	5132	5145	5159	5172	1.3
33	5185	5198	5211	5224	5237	5250	5263	5276	5289	5302	1.3
34	5315	5328	5340	5353	5366	5378	5391	5403	5416	5428	1.3
35	5441	5453	5465	5478	5490	5502	5514	5527	5539	5551	1.2
36	5563	5575	5587	5599	5611	5623	5635	5647	5658	5670	1.2
37	5682	5694	5705	5717	5729	5740	5752	5763	5775	5786	1.2
38	5798	5809	5821	5832	5843	5855	5866	5877	5888	5899	1.2
39	5911	5922	5933	5944	5955	5966	5977	5988	5999	6010	1.2
40	6021	6031	6042	6053	6064	6075	6085	6096	6107	6117	1.2
41	6128	6138	6149	6160	6170	6180	6191	6201	6212	6222	1.2
42	6232	6243	6253	6263	6274	6284	6294	6304	6314	6325	1.2
43	6335	6345	6355	6365	6375	6385	6395	6405	6415	6425	1.2
44	6435	6444	6454	6464	6474	6484	6493	6503	6513	6522	1.2

MENSURATION.

N	0	1	2	3	4	5	6	7	8	9	PP
45	6532	6542	6551	6561	6571	6580	6590	6599	6609	6618	1.0
46	6628	6637	6646	6656	6665	6675	6684	6693	6702	6712	1.0
47	6721	6730	6739	6749	6758	6767	6776	6785	6794	6803	1.0
48	6812	6821	6830	6839	6848	6857	6866	6875	6884	6893	1.0
49	6902	6911	6920	6928	6937	6946	6955	6964	6972	6981	1.0
50	6990	6998	7007	7016	7024	7033	7042	7050	7059	7067	1.0
51	7076	7084	7093	7101	7110	7118	7126	7135	7143	7152	1.0
52	7160	7168	7177	7185	7193	7202	7210	7218	7226	7235	1.0
53	7243	7251	7259	7267	7275	7284	7292	7300	7308	7316	1.0
54	7324	7332	7340	7348	7356	7364	7372	7380	7388	7396	1.0
55	7404	7412	7419	7427	7435	7443	7451	7459	7466	7474	0.8
56	7482	7490	7497	7505	7513	7520	7528	7536	7543	7551	0.8
57	7559	7566	7574	7582	7589	7597	7604	7612	7619	7627	0.8
58	7634	7642	7649	7657	7664	7672	7679	7686	7694	7701	0.7
59	7709	7716	7723	7731	7738	7745	7752	7760	7767	7774	0.7
60	7782	7789	7796	7803	7810	7818	7825	7832	7839	7846	0.7
61	7853	7860	7868	7875	7882	7889	7896	7903	7910	7917	0.7
62	7924	7931	7938	7945	7952	7959	7966	7973	7980	7987	0.7
63	7993	8000	8007	8014	8021	8028	8035	8041	8048	8055	0.7
64	8062	8069	8075	8082	8089	8096	8102	8109	8116	8122	0.7
65	8129	8136	8142	8149	8156	8162	8169	8176	8182	8189	0.7
66	8195	8202	8209	8215	8222	8228	8235	8241	8248	8254	0.7
67	8261	8267	8274	8280	8287	8293	8299	8306	8312	8319	0.6
68	8325	8331	8338	8344	8351	8357	8363	8370	8376	8382	0.6
69	8388	8395	8401	8407	8414	8420	8426	8432	8439	8445	0.6
70	8451	8457	8463	8470	8476	8482	8488	8494	8500	8506	0.6
71	8513	8519	8525	8531	8537	8543	8549	8555	8561	8567	0.6
72	8573	8579	8585	8591	8597	8603	8609	8615	8621	8627	0.6
73	8633	8639	8645	8651	8657	8663	8669	8675	8681	8686	0.6
74	8692	8698	8704	8710	8716	8722	8727	8733	8739	8745	0.6
75	8751	8756	8762	8768	8774	8779	8785	8791	8797	8802	0.6
76	8808	8814	8820	8825	8831	8837	8842	8848	8854	8859	0.6
77	8865	8871	8876	8882	8887	8893	8899	8904	8910	8915	0.6
78	8921	8927	8932	8938	8943	8949	8954	8960	8965	8971	0.6
79	8976	8982	8987	8993	8998	9004	9009	9015	9020	9025	0.5

LOGARITHMS.

N	0	1	2	3	4	5	6	7	8	9	PP
80	9031	9036	9042	9047	9053	9058	9063	9069	9074	9079	0.5
81	9085	9090	9096	9101	9106	9112	9117	9122	9128	9133	0.5
82	9138	9143	9149	9154	9159	9165	9170	9175	9180	9186	0.5
83	9191	9196	9201	9206	9212	9217	9222	9227	9232	9238	0.5
84	9243	9248	9253	9258	9263	9269	9274	9279	9284	9289	0.5
85	9294	9299	9304	9309	9315	9320	9325	9330	9335	9340	0.5
86	9345	9350	9355	9360	9365	9370	9375	9380	9385	9390	0.5
87	9395	9400	9405	9410	9415	9420	9425	9430	9435	9440	0.5
88	9445	9450	9455	9460	9465	9469	9474	9479	9484	9489	0.5
89	9494	9499	9504	9509	9513	9518	9523	9528	9533	9538	0.5
90	9542	9547	9552	9557	9562	9566	9571	9576	9581	9586	0.5
91	9590	9595	9600	9605	9609	9614	9619	9624	9628	9633	0.5
92	9638	9643	9647	9652	9657	9661	9666	9671	9675	9680	0.5
93	9685	9689	9694	9699	9703	9708	9713	9717	9722	9727	0.5
94	9731	9736	9741	9745	9750	9754	9759	9763	9768	9773	0.5
95	9777	9782	9786	9791	9795	9800	9805	9809	9814	9818	0.5
96	9823	9827	9832	9836	9841	9845	9850	9854	9859	9863	0.5
97	9868	9872	9877	9881	9886	9890	9894	9899	9903	9908	0.4
98	9912	9917	9921	9926	9930	9934	9939	9943	9948	9952	0.4
99	9956	9961	9965	9969	9974	9978	9983	9987	9991	9996	0.4

N	0	1	2	3	4	5	6	7	8	9
100	0000	0004	0009	0013	0017	0022	0026	0030	0035	0039
101	0043	0048	0052	0056	0060	0065	0069	0073	0077	0082
102	0086	0090	0095	0099	0103	0107	0111	0116	0120	0124
103	0128	0133	0137	0141	0145	0149	0154	0158	0162	0166
104	0170	0175	0179	0183	0187	0191	0195	0199	0204	0208
105	0212	0216	0220	0224	0228	0233	0237	0241	0245	0249
106	0253	0257	0261	0265	0269	0273	0278	0282	0286	0290
107	0294	0298	0302	0306	0310	0314	0318	0322	0326	0330
108	0334	0338	0342	0346	0350	0354	0358	0362	0366	0370
109	0374	0378	0382	0386	0390	0394	0398	0402	0406	0410

230 MENSURATION.

N	0	1	2	3	4	5	6	7	8	9
110	0414	0418	0422	0426	0430	0434	0438	0441	0445	0449
111	0453	0457	0461	0465	0469	0473	0477	0481	0484	0488
112	0492	0496	0500	0504	0508	0512	0515	0519	0523	0527
113	0531	0535	0538	0542	0546	0550	0554	0558	0561	0565
114	0569	0573	0577	0580	0584	0588	0592	0596	0599	0603
115	0607	0611	0615	0618	0622	0626	0630	0633	0637	0641
116	0645	0648	0652	0656	0660	0663	0667	0671	0674	0678
117	0682	0686	0689	0693	0697	0700	0704	0708	0711	0715
118	0719	0722	0726	0730	0734	0737	0741	0745	0748	0752
119	0755	0759	0763	0766	0770	0774	0777	0781	0785	0788
120	0792	0795	0799	0803	0806	0810	0813	0817	0821	0824
121	0828	0831	0835	0839	0842	0846	0849	0853	0856	0860
122	0864	0867	0871	0874	0878	0881	0885	0888	0892	0896
123	0899	0903	0906	0910	0913	0917	0920	0924	0927	0931
124	0934	0938	0941	0945	0948	0952	0955	0959	0962	0966
125	0969	0973	0976	0980	0983	0986	0990	0993	0997	1000
126	1004	1007	1011	1014	1017	1021	1024	1028	1031	1035
127	1038	1041	1045	1048	1052	1055	1059	1062	1065	1069
128	1072	1075	1079	1082	1086	1089	1092	1096	1099	1103
129	1106	1109	1113	1116	1119	1123	1126	1129	1133	1136
130	1139	1143	1146	1149	1153	1156	1159	1163	1166	1169
131	1173	1176	1179	1183	1186	1189	1193	1196	1199	1202
132	1206	1209	1212	1216	1219	1222	1225	1229	1232	1235
133	1239	1242	1245	1248	1252	1255	1258	1261	1265	1268
134	1271	1274	1278	1281	1284	1287	1290	1294	1297	1300
135	1303	1307	1310	1313	1316	1319	1323	1326	1329	1332
136	1335	1339	1342	1345	1348	1351	1355	1358	1361	1364
137	1367	1370	1374	1377	1380	1383	1386	1389	1392	1396
138	1399	1402	1405	1408	1411	1414	1418	1421	1424	1427
139	1430	1433	1436	1440	1443	1446	1449	1452	1455	1458
140	1461	1464	1467	1471	1474	1477	1480	1483	1486	1489
141	1492	1495	1498	1501	1504	1508	1511	1514	1517	1520
142	1523	1526	1529	1532	1535	1538	1541	1544	1547	1550
143	1553	1556	1559	1562	1565	1569	1572	1575	1578	1581
144	1584	1587	1590	1593	1596	1599	1602	1605	1608	1611

LOGARITHMS. 231

N	0	1	2	3	4	5	6	7	8	9
145	1614	1617	1620	1623	1626	1629	1632	1635	1638	1641
146	1644	1647	1649	1652	1655	1658	1661	1664	1667	1670
147	1673	1676	1679	1682	1685	1688	1691	1694	1697	1700
148	1703	1706	1708	1711	1714	1717	1720	1723	1726	1729
149	1732	1735	1738	1741	1744	1746	1749	1752	1755	1758
150	1761	1764	1767	1770	1772	1775	1778	1781	1784	1787
151	1790	1793	1796	1798	1801	1804	1807	1810	1813	1816
152	1818	1821	1824	1827	1830	1833	1836	1838	1841	1844
153	1847	1850	1853	1855	1858	1861	1864	1867	1870	1872
154	1875	1878	1881	1884	1886	1889	1892	1895	1898	1901
155	1903	1906	1909	1912	1915	1917	1920	1923	1926	1928
156	1931	1934	1937	1940	1942	1945	1948	1951	1953	1956
157	1959	1962	1965	1967	1970	1973	1976	1978	1981	1984
158	1987	1989	1992	1995	1998	2000	2003	2006	2009	2011
159	2014	2017	2019	2022	2025	2028	2030	2033	2036	2038
160	2041	2044	2047	2049	2052	2055	2057	2060	2063	2066
161	2068	2071	2074	2076	2079	2082	2084	2087	2090	2092
162	2095	2098	2101	2103	2106	2109	2111	2114	2117	2119
163	2122	2125	2127	2130	2133	2135	2138	2140	2143	2146
164	2148	2151	2154	2156	2159	2162	2164	2167	2170	2172
165	2175	2177	2180	2183	2185	2188	2191	2193	2196	2198
166	2201	2204	2206	2209	2212	2214	2217	2219	2222	2225
167	2227	2230	2232	2235	2238	2240	2243	2245	2248	2251
168	2253	2256	2258	2261	2263	2266	2269	2271	2274	2276
169	2279	2281	2284	2287	2289	2292	2294	2297	2299	2302
170	2304	2307	2310	2312	2315	2317	2320	2322	2325	2327
171	2330	2333	2335	2338	2340	2343	2345	2348	2350	2353
172	2355	2358	2360	2363	2365	2368	2370	2373	2375	2378
173	2380	2383	2385	2388	2390	2393	2395	2398	2400	2403
174	2405	2408	2410	2413	2415	2418	2420	2423	2425	2428
175	2430	2433	2435	2438	2440	2443	2445	2448	2450	2453
176	2455	2458	2460	2463	2465	2467	2470	2472	2475	2477
177	2480	2482	2485	2487	2490	2492	2494	2497	2499	2502
178	2504	2507	2509	2512	2514	2516	2519	2521	2524	2526
179	2529	2531	2533	2536	2538	2541	2543	2545	2548	2550

MENSURATION.

N	0	1	2	3	4	5	6	7	8	9
180	2553	2555	2558	2560	2562	2565	2567	2570	2572	2574
181	2577	2579	2582	2584	2586	2589	2591	2594	2596	2598
182	2601	2603	2605	2608	2610	2613	2615	2617	2620	2622
183	2625	2627	2629	2632	2634	2636	2639	2641	2643	2646
184	2648	2651	2653	2655	2658	2660	2662	2665	2667	2669
185	2672	2674	2676	2679	2681	2683	2686	2688	2690	2693
186	2695	2697	2700	2702	2704	2707	2709	2711	2714	2716
187	2718	2721	2723	2725	2728	2730	2732	2735	2737	2739
188	2742	2744	2746	2749	2751	2753	2755	2758	2760	2762
189	2765	2767	2769	2772	2774	2776	2778	2781	2783	2785
190	2788	2790	2792	2794	2797	2799	2801	2804	2806	2808
191	2810	2813	2815	2817	2819	2822	2824	2826	2828	2831
192	2833	2835	2838	2840	2842	2844	2847	2849	2851	2853
193	2856	2858	2860	2862	2865	2867	2869	2871	2874	2876
194	2878	2880	2882	2885	2887	2889	2891	2894	2896	2898
195	2900	2903	2905	2907	2909	2911	2914	2916	2918	2920
196	2923	2925	2927	2929	2931	2934	2936	2938	2940	2942
197	2945	2947	2949	2951	2953	2956	2958	2960	2962	2964
198	2967	2969	2971	2973	2975	2978	2980	2982	2984	2986
199	2989	2991	2993	2995	2997	2999	3002	3004	3006	3008

www.ingramcontent.com/pod-product-compliance
Lightning Source LLC
Chambersburg PA
CBHW020804230426
43666CB00007B/852